Addressing Special Educational Needs and Disability in the Curriculum: Design and Technology

The SEND Code of Practice (2015) reinforced the requirement that all teachers must meet the needs of all learners. This topical book provides practical, tried-and-tested strategies and resources that will support teachers in making design and technology lessons accessible and interesting for all pupils, including those with special educational needs. The author draws on a wealth of experience to share her understanding of special educational needs and disabilities and show how the design and technology teacher can reduce or remove any barriers to learning.

Offering strategies that are specific to the context of design and technology teaching, this book will enable teachers to:

- better identify a student's particular learning requirements;
- set inclusive design and making assignments which allow all students to participate and succeed;
- build students' confidence in using a range of materials and tools;
- assist with design tasks where pupils take ownership of their work and learning;
- adapt the classroom environment to meet the needs of pupils;
- create a mutually supportive classroom which maximises learning opportunities.

An invaluable tool for continuing professional development, this text will be essential for design and technology teachers (and their teaching assistants) seeking to include and motivate all pupils in their lessons, regardless of their individual needs. This book will also be of interest to secondary SENCOs, senior management teams and ITT providers.

In addition to free online resources, a range of appendices provide design and technology teachers with a variety of pro forma and activity sheets to support effective teaching. This is an essential tool for design and technology teachers and teaching assistants, and will help to deliver successful, inclusive lessons for all pupils.

Louise T. Davies is founder of the Food Teachers Centre and advisor to the Department for Education.

Addressing Special Educational Needs and Disability in the Curriculum

Series editor: Linda Evans

Children and young people with a diverse range of special educational needs and disabilities (SEND) are expected to access the full curriculum. Crucially, the current professional standards make it clear that *every* teacher must take responsibility for *all* pupils in their classes. Titles in this fully revised and updated series will be essential for teachers seeking subject-specific guidance on meeting their pupils' individual needs. In line with recent curriculum changes, the new Code of Practice for SEN and other pedagogical developments, these titles provide clear, practical strategies and resources that have proved to be effective and successful in their particular subject area. Written by practitioners, they can be used by departmental teams and in 'whole-school' training sessions as professional development resources. With free Web-based online resources also available to complement the books, these resources will be an asset to any teaching professional helping to develop policy and provision for learners with SEND.

The new national curriculum content will prove challenging for many learners, and teachers of children in Y5 and Y6 will also find the books a valuable resource.

Titles in this series include:

Addressing Special Educational Needs and Disability in the Curriculum: Modern Foreign Languages
John Connor

Addressing Special Educational Needs and Disability in the Curriculum: Music
Victoria Jaquiss and Diane Paterson

Addressing Special Educational Needs and Disability in the Curriculum: PE and Sport
Crispin Andrews

Addressing Special Educational Needs and Disability in the Curriculum: Science
Marion Frankland

Addressing Special Educational Needs and Disability in the Curriculum: Design and Technology
Louise T. Davies

Addressing Special Educational Needs and Disability in the Curriculum: History
Ian Luff and Richard Harris

Addressing Special Educational Needs and Disability in the Curriculum: Religious Education
Dilwyn Hunt

Addressing Special Educational Needs and Disability in the Curriculum: Geography
Graeme Eyre

Addressing Special Educational Needs and Disability in the Curriculum: Art
Gill Curry and Kim Earle

Addressing Special Educational Needs and Disability in the Curriculum: English
Tim Hurst

Addressing Special Educational Needs and Disability in the Curriculum: Maths
Max Wallace

Addressing Special Educational Needs and Disability in the Curriculum: Design and Technology

Second edition

Louise T. Davies

 Routledge
Taylor & Francis Group

LONDON AND NEW YORK

Second edition published 2018
by Routledge
2 Park Square, Milton Park, Abingdon, Oxon OX14 4RN

and by Routledge
711 Third Avenue, New York, NY 10017

Routledge is an imprint of the Taylor & Francis Group, an informa business

First edition published 2004 by David Fulton Publishers as *Meeting SEN in the Curriculum: Design and Technology* by Louise T. Davies and Elisabeth Barratt-Hacking

British Library Cataloguing in Publication Data
A catalogue record for this book is available from the British Library

Library of Congress Cataloging in Publication Data
A catalog record for this book has been requested

ISBN: 978-1-138-71492-2 (hbk)
ISBN: 978-0-415-37685-3 (pbk)
ISBN: 978-1-315-23110-5 (ebk)

Typeset in Helvetica
by Keystroke, Neville Lodge, Tettenhall, Wolverhampton

Visit eResources: www.routledge.com/9781138714922

Contents

Appendices

Series authors

The author

Louise T. Davies offers advice and guidance to the DfE and key food education organisations based on her years of experience as a teacher and teacher trainer, and in curriculum development at QCA and the Royal College of Art. From 2005–13 she led innovation at the Design & Technology Association and provided teaching expertise for a diverse range of high-profile D&T curriculum and professional development programmes. She has provided specialist advice on teaching standards and best practice; teacher training and CPD; special educational needs; food and textiles education (Food in Schools programmes, Licence to Cook, Teach Food Technology, Active Kids Get Cooking Teaching Food Safely); GCSE Textiles Rescue programme for lower attaining schools; digital technologies. Louise has also worked with the James Dyson Foundation Innovation Group for outstanding teachers.

She's a prolific author, having written over fifty D&T textbooks and award-winning multimedia teaching resources. Most recently, she worked as lead consultant for the School Food Champions Programme (2013–16) and as an adviser to the DfE on the new GCSE in food preparation and nutrition.

In 2013, Louise founded the Food Teachers Centre, a UK-based self-help group supported by experienced associates. It provides a platform to exchange best practice, give advice and support to less experienced teachers, answering practical concerns and keeping them abreast of the latest curriculum changes. It hosts regional and national events, which reach around 2,000 secondary food teachers each year. The centre works with all the major government agencies (Ofsted, DfE, the School Food Plan) and with key industry representatives (British Nutrition Foundation, Agriculture Horticulture Development Board, Chilled Food Association, Institute of Food Science and Technology, Springboard UK) to promote its aim of better food teaching.

Series editor

Linda Evans was commissioning editor for the original books in this series and has co-ordinated the updating process for these new editions. She has taught children of all ages over the years and posts have included those of SENCO, LA Adviser, Ofsted inspector and HE tutor/lecturer. She was awarded a PhD in 2000 following research on improving educational outcomes for children (primary and secondary).Since then, Linda has been commissioning editor for David Fulton Publishing (SEN) as well as editor of a number of educational journals and newsletters: she has also written books, practical classroom resources, Masters course materials and school improvement guidance. She maintains her contact with school practitioners through her work as a part-time ITT tutor and educational consultant.

SEND specialist

Sue Briggs has been supporting the education and inclusion of children with special educational needs and disabilities, and their parents, for over 20 years; variously as teacher, Ofsted inspector, specialist member of the SEN and Disability Tribunal, school improvement partner, consultant and adviser. She holds a Masters degree in education, a first class BEd and a diploma in special education (DPSE distinction). Sue was a national lead for the Achievement for All programme (2011–13) and a regional adviser for the Early Support programme for the Council for Disabled Children (2014–15) and is currently an independent education and leadership consultant.

Sue is the author of several specialist books and publications including *Meeting SEND in Primary Classrooms* and *Meeting SEND in Secondary Classrooms* (Routledge, 2015).

Subject specialists

Art

Gill Curry was Head of Art in a secondary school in Wirral for 20 years and advisory teacher for art and gifted and talented strand coordinator. She has an MA in print from the University of Chester and an MA in women's studies from the University of Liverpool.

She is a practising artist, specialising in print, and exhibits nationally and internationally, running courses regularly in schools and print studios.

Kim Earle is vice principal at Birkenhead High School Academy for Girls on the Wirral. She has previously been a head of art and head of creative arts, securing Artsmark Gold in all the establishments in which she has worked. Kim was also formerly Able Pupils and Arts Consultant in St Helens, working across special schools and mainstream schools with teaching and support staff on art policy and practice. She still teaches art in a mixed ability setting in her current school and works closely with local schools and outside organisations to address barriers to learning.

English

Tim Hurst began his career as an English teacher at the Willian School in Hertfordshire, becoming Second in English before deciding that his future lay in SEND. He studied for an advanced diploma in special educational needs and has been a SEN coordinator in five schools in Hertfordshire, Essex and Suffolk. Tim has always been committed to the concept of inclusion and is particularly interested in reading development, which he passionately believes in as a whole-school responsibility.

Geography

Graeme Eyre has considerable experience of teaching and leading geography in secondary schools in a range of different contexts, and is currently Assistant Principal for Intervention at an academy in inner London. Graeme is a consultant to the Geographical Association and a Fellow of the Royal Geographical Society. He has also delivered training and CPD for teachers at all levels. He holds a BA in geography, a PGCE in secondary geography and an MA in geography education.

History

Ian Luff taught in comprehensive schools for 32 years and was head of history in four such schools, writing extensively and delivering training in teaching the subject. He served in the London Borough of Barking and Dagenham as advisory teacher and as deputy headteacher at Kesgrave High School in Suffolk. Ian was made an honorary fellow of the Historical Association for contributions to education in 2011 and is currently an associate tutor and PhD student in the School of Education and Lifelong Learning at the University of East Anglia.

Richard Harris taught in comprehensive schools for 16 years, and was a head of history and head of humanities, as well as teacher consultant for history in West Berkshire. He has spent 15 years working with trainee history teachers at the universities of Southampton and Reading and is currently director of teaching and learning as well as researching issues mainly relating to history education. He has advised government bodies and worked extensively with the Council of Europe on teacher education and history education. He was made an honorary fellow of the Historical Association in 2011.

Languages

John Connor is a former head of faculty, local authority adviser and senior examiner. He has served as an Ofsted team inspector for modern languages and special educational needs in mainstream settings. John was also an assessor on the Advanced Skills Teacher programme for the DfE. He is currently working as a trainer, author and consultant, and has directed teaching and learning quality audits across England, the Channel Islands, Europe, the Middle East and the Far East. He is also a governor of a local primary school.

Maths

Max Wallace has nine years' experience of teaching children with special educational needs. He currently works as an advanced skills teacher at an inclusive mainstream secondary school. Appointed as a specialist leader in education for mathematics, Max mentors and coaches teachers in a wide network of schools. He has previously worked as a head of year and was responsible for the continuing professional development of colleagues. He has a doctorate in mathematics from Cardiff University.

Music

Victoria Jaquiss (FRSA) trained as a teacher of English and Drama and held posts of English teacher, Head of PSE, Music and Expressive Arts at Foxwood School. She became a recognised authority on behaviour management and inclusion with children in challenging circumstances. The second half of her career has involved working for the Leeds Music Service/Leeds ArtForms as Steel Pan Development Officer and deputy inclusion manager/teacher. She was awarded the fellowship of the Royal Society of Arts in 2002.

Diane Paterson began teaching as a mainstream secondary music teacher. She went on to study how music technology could enable people with severe physical difficulties to make their own music, joining the Drake Music project in Yorkshire and becoming its regional leader. She then became inclusion manager/teacher at Leeds Music Service/ArtForms, working with children with additional needs. As secretary of YAMSEN: SpeciallyMusic, she now runs specialist regional workshops, music days and concerts for students with special/additional needs and their carers.

PE and sport

Crispin Andrews is a qualified teacher and sports coach, and has worked extensively in Buckinghamshire schools coaching cricket and football and developing opportunities for girls in these two sports. He is currently a sports journalist, writing extensively for a wide range of educational journals, including *Special Children* and the *Times Educational Supplement*, and other publications such as *Cricket World*.

Religious education

Dilwyn Hunt taught RE for 18 years before becoming an adviser, first in Birmingham and then in Dudley. He currently works as an independent RE adviser supporting local authorities, SACREs and schools. He is also in demand across the country as a speaker on all aspects of teaching RE, in both mainstream and special settings. He is the author of numerous popular classroom resources and books and currently serves as the executive assistant on the Association of RE Inspectors, Advisers and Consultants.

Science

Marion Frankland (CSciTeach) has been teaching for 16 years and was an advanced skills teacher of science. She has extensive experience of teaching science at all levels in both mainstream and special schools, and has worked as a SENCO in a special school, gaining her qualification alongside her teaching commitment.

A few words from the series editor

The original version of this book formed part of the 'Meeting SEN in the Curriculum' series which was published ten years ago to much acclaim. The series won a BERA (British Educational Resources Award) and has been widely used by ITT providers, their students and trainees, curriculum and SEN advisers, department heads and teachers of all levels of experience. It has proven to be highly successful in helping to develop policy and provision for learners with special educational needs or disabilities.

The series was born out of an understanding that practitioners want information and guidance about improving teaching and learning that is *relevant to them* – rooted in their particular subject, and applicable to pupils they encounter. These books exactly fulfil that function.

Those original books have stood the test of time in many ways – their tried and tested, practical strategies are as relevant and effective as ever. Legislation and national guidance has moved on however, as have resources and technology; new terminology accompanies all of these changes. For example, we have changed the series title to incorporate the acronym 'SEND' (special educational needs or disability), which has been adopted in official documents and in many schools in response to recent legislation and the revised Code of Practice. The important point to make is that our authors have addressed the needs of pupils with a wide range of special or 'additional' needs; some will have educational, health and care (EHC) plans which have replaced 'statements', but most will not. Some will have identified 'syndromes' or 'conditions' but many will simply be termed 'low attainers'; pupils who, for whatever reason, do not easily make progress.

This second edition encompasses recent developments in education, and specifically in the teaching of design and technology. At the time of publication, education is still very much in an era of change; our national curriculum, monitoring and assessment systems are all newly fashioned and many schools are still adjusting to changes and developing their own ways forward. The ideas and guidance contained in this book, however, transcend the fluctuations of national politics and policy and provide a framework for ensuring that pupils with SEND can 'enjoy and achieve' in their D&T lessons.

NB: The term 'parent' is used throughout and is intended to cover any adult who is a child's main care-giver.

Linda D. Evans

Acknowledgements

This book is dedicated to the very special staff and pupils at Aspen House School, London. Without you, I would have left teaching in 1986 to sell microwave ovens. How much deeper you made my world!

I am indebted to many teachers and colleagues from special schools all over the UK, who have helped me over many years towards the first and second editions of this book. These include teachers who worked with me at the Royal College of Art Schools Technology Project, Qualifications and Curriculum Authority (QCA/QCDA) and the Design and Technology Association (Special Needs Advisory Group). I would also like to thank those teachers who shared ideas and resources at our 2016 'Food is Special' event, to help to update the resources for this edition.

The many focus group meetings, visits to schools, training events and coaching sessions showed me how extraordinary you all are at adapting D&T activities for your pupils: we have a lot to learn from you about differentiation and individualised learning.

Thanks to those people who allowed me to describe their teaching activities in the first and second editions, especially Steve Gilbert, Brian Russell, Hugh Sammons, Sue Stalley, Heather Wither, Donna Trebell, Chris Killey, Chris Clarke, Josie Brown, Gary Drabble and Mike Lewis.

Particular thanks in this edition are due to Becky Wells at Sutherland House School (for the example of how to adapt the 4 × 4 designing activity for pupils with ASD), Rachel Gregory (for the photo of yeast experiments with balloons), and Debbie Green and John Jamieson School, Leeds (for the puppets designing example).

Photo credits

The author would like to thank the following for kindly granting permission to reproduce their content in this book:

Roy Ballam, Louise T. Davies and Nikki Young for the Technopacks – Chapter 3 SEN Design Sheets, key words sheets and certificate.

Becky Wells and Debbie Green for designing examples (pages 45 and 46)

David Evans of Fox Lane Photography.

Introduction

Ours to teach

Your class: busy designing and making practical lessons with a wide range of individuals to teach – to encourage, motivate and inspire; all individuals who must be seen to make good progress regardless of their various abilities, backgrounds, interests and personalities. This is what makes teaching so interesting!

Jason demonstrates very little interest in school. He rarely completes homework and frequently turns up without a pen. He finds it hard to listen when you're talking and is likely to start his own conversation with a classmate. His work is untidy and mostly incomplete. It's difficult to find evidence of his progress this year.

Zoe tries very hard in lessons but is slow to understand explanations and has difficulty in expressing herself. She has been assessed as having poor communication skills but there is no additional resourcing for her.

Ethan is on the autistic spectrum and finds it difficult to relate to other people, to work in a group and to understand social norms. He has an education, health and care plan which provides for some TA support but this is not timetabled for all lessons.

Do you recognise these youngsters? Our school population is now more diverse than ever before, with pupils of very different abilities, aptitudes and interests, from a wide range of cultures making up our mainstream and special school classes. Many of these learners will experience difficulties of some sort at school, especially when they are faced with higher academic expectations at the end of KS2 and into KS3–4.

Whether they have a specific special educational need like dyslexia, or are on the autistic spectrum, or for various reasons cannot conform to our behavioural expectations – *they are ours to teach*. Our lessons must ensure that each and every pupil can develop their skills and knowledge and make good progress.

How can this book help?

The information, ideas and guidance in this book will enable teachers of design and technology (and their teaching assistants) to plan and deliver lessons that will meet the individual needs of learners who experience difficulties. It will be especially valuable to subject teachers because the ideas and guidance are provided within a subject context, ensuring relevance and practicability.

Teachers who cater well for pupils with special educational needs and disabilities (SEND) are likely to cater well for *all* pupils – demonstrating outstanding practice in their everyday teaching. These teachers have a keen awareness of the many factors affecting a pupil's ability to learn, not only characteristics of the individual but also aspects of the learning environment that can either help or hinder learning. This book will help practitioners to develop strategies that can be used selectively to enable each and every learner to make progress.

Professional development

Our education system is constantly changing. The national curriculum, SEND legislation, examination reform and significant change to Ofsted inspection means that teachers need to keep up to date and be able to develop the knowledge, skills and understanding necessary to meet the needs of all the learners they teach. High quality continuing professional development (CPD) has a big part to play in this.

Faculties and subject teams planning for outstanding teaching and learning should consider how they regularly review and improve their provision by:

* auditing

   a) the skills and expertise of current staff (teachers and assistants);
   b) their professional development needs for SEND, based on the current cohorts of pupils;

   (An audit proforma can be found in the eResources at: www.routledge.com/9781138714922)

* using the information from the two audits to develop a CPD programme (using internal staff, colleagues from nearby schools and/or consultants to deliver bespoke training);

- enabling teachers to observe each other, teach together, visit other class-rooms and other schools;
- encouraging staff to reflect on their practice and feel comfortable in sharing both the positive and the negative experiences;
- establishing an ethos that values everyone's expertise (including pupils and parents who might be able to contribute to training sessions);
- using online resources that are readily available to support workforce development (e.g.www.nasen.org.uk/onlinesendcpd/);
- encouraging staff to access (and disseminate) further study and high quality professional development.

This book, and the others in the series, will be invaluable in contributing to whole-school CPD on meeting special educational needs, and in facilitating subject-specific staff development within departments.

1 Meeting special educational needs and disabilities

Your responsibility

New legislation and national guidance in 2014 changed the landscape of educational provision for pupils with any sort of 'additional' or 'special' needs. The vast majority of learners, including those with 'moderate' or 'mild' learning difficulties, weak communication skills, dyslexia or social/behavioural needs rarely attract additional resources: they are very much accepted as part of the 'mainstream mix'. Pupils with more significant special educational needs and/or disabilities (SEND) may have an education, health and care plan (EHC plan): this outlines how particular needs will be met, often involving professionals from different disciplines, and sometimes specifying adult support in the classroom. Both groups of pupils are ultimately the responsibility of the class teacher, whether in mainstream or special education.

> High quality teaching that is differentiated and personalised will meet the individual needs of the majority of children and young people. Some children and young people need educational provision that is additional to or different from this. This is special educational provision under Section 21 of the Children and Families Act 2014. Schools and colleges *must* use their best endeavours to ensure that such provision is made for those who need it. Special educational provision is underpinned by high quality teaching and is compromised by anything less.
>
> (SEND Code of Practice: DfE 2015)

There is more information about legislation (The Children and Families Act 2014; The Equality Act 2010) and guidance (SEND Code of Practice) in Appendix 1.1.

Definition of SEND

A pupil has special educational needs if he or she:

- has a significantly greater difficulty in learning than the majority of others of the same age; or
- has a disability which prevents or hinders him or her from making use of facilities of a kind generally provided for others of the same age in mainstream schools or mainstream Post-16 institutions.

(SEND Code of Practice: DfE 2015)

The SEND Code of Practice identifies four broad areas of SEND (see Table 1.1), but remember that this gives only an overview of the range of needs that should be planned for by schools; pupils' needs rarely fit neatly into one area of need.

Whole-school ethos

Successful schools are pro-active in identifying and addressing pupils' special needs, focusing on adapting the educational context and environment rather than on 'fixing' an individual learner. Adapting systems and teaching programmes rather than trying to force the pupil to conform to rigid expectations will lead to a greater chance of success in terms of learning outcomes.

Table 1.1 The four broad areas of SEND

Communication and interaction	Cognition and learning	Social, emotional and mental health difficulties	Sensory and/or physical needs
Speech, language and communication needs (SLCN)	Specific learning difficulties (SpLD)	Mental health difficulties such as anxiety or depression, self-harming, substance abuse or eating disorders	Vision impairment (VI)
Asperger's Syndrome and Autism (ASD)	Moderate learning difficulties (MLD)		Hearing impairment (HI)
	Severe learning difficulties (SLD)	Attention deficit disorders, attention deficit hyperactivity disorder or attachment disorder	Multi-sensory impairment (MSI)
	Profound and multiple learning difficulties (PMLD)		Physical disability (PD)

Guidance on whole-school and departmental policy making can be found in Appendix 1.2 and a sample D&T policy for SEND can be downloaded from www.routledge.com/9781138714922.

Policy into practice

In many cases, pupils' individual learning needs will be met through differentiation of tasks and materials in their lessons; sometimes this will be supplemented by targeted interventions such as 'catch-up' programmes delivered outside the classroom. A smaller number of pupils may need access to more specialist equipment and approaches, perhaps based on advice and support from external specialists.

The main thrust of the Children and Families Act and Chapter 6 of the SEND Code of Practice is that outcomes for pupils with SEN must be improved and that schools and individual teachers must have high aspirations and expectations for all.

In practice, this means that pupils should be enabled to:

- **achieve their best**; additional provision made for pupils with SEND will enable them to make accelerated progress so that the gap in progress and attainment between them and other pupils is reduced. Being identified with SEND should no longer be a reason for a pupil making less than good progress.
- **become confident individuals living fulfilling lives**; if you ask parents of children with SEND what is important to them for their child's future they often answer 'happiness, the opportunity to achieve his or her potential, friendships, and a loving family' – just what we all want for our children. Outcomes in terms of well-being, social skills and growing independence are equally as important as academic outcomes for children and young people with SEND. D&T is an important vehicle for developing practical capabilities for everyday life, such as being able to make things, cook and follow a plan.
- **make a successful transition into adulthood, whether into employment, further or higher education or training;** decisions made at transition from primary school, in Year 7 and beyond should be made in the context of preparation for adulthood. For example, where a pupil has had full-time support from a teaching assistant in primary school, the secondary school's first reaction might be to continue this level of support after transition. This may result in long-term dependency on adults however, or limited opportunities to develop social skills, both of which impact negatively on preparation for adulthood. Planning a D&T experience appropriate for pupils with SEND involves developing confident young people, as they

have had an opportunity to practice work and life-related skills, who are becoming independent makers and cope well with real life decisions and choices. Opportunities for enterprise and craftsmanship are very important.

Excellent classroom provision

Later chapters provide lots of subject-specific ideas and guidance on strategies to support pupils with SEND. In Chapter 2 you will find useful guidelines to help you support pupils with identified 'conditions' but there are some generic approaches that form the foundations of outstanding provision, such as:

- providing support from adults or other pupils;
- adapting tasks or environments;
- using specialist aids and equipment as appropriate.

The starting points listed below provide a sound basis for creating an inclusive learning environment that will benefit *all* pupils, while being especially important for those with SEND.

Develop pupils' understanding through the use of all available senses by:

- using 3D resources that pupils can access through sight *and* sound (the senses of touch, taste and smell in D&T will broaden understanding and ensure stronger memory);
- regularly employing resources such as symbols, pictures and film to increase pupils' knowledge of the wider world and contextualise new information and skills;
- encouraging and enabling pupils to take part in activities such as exploring or playing with everyday objects; informally and formally evaluating and assessing consumer products; production simulation and enterprise activities, class visits and exploring the environment.

Help pupils to learn effectively and prepare for further or higher education, work or training by:

- setting realistic demands within high expectations;
- using positive strategies to manage behaviour;
- giving pupils opportunities and encouragement to develop the skills to work effectively in a group or with a partner;
- teaching all pupils to value and respect the contribution of others;
- encouraging independent working skills;
- teaching essential safety rules.

Help pupils to develop communication skills, language and literacy by:

- making sure all pupils can see your face when you are speaking;
- giving clear, step-by-step instructions, and limiting the amount of information given at one time;
- providing a list of key vocabulary for each lesson;
- choosing texts that pupils can read and understand;
- making texts available in different formats, including large text, symbols or by using screen reader programs;
- putting headings and important points in bold or highlighting to make them easier to scan;
- presenting written information as concisely as possible, using bullet points, images or diagrams.

Support pupils with disabilities by:

- encouraging pupils to be as independent as possible;
- enabling them to work with other, non-disabled pupils;
- making sure the classroom environment is suitable, e.g. uncluttered space to facilitate movement around the classroom, food room or workshop; adapted resources that are labeled and accessible;
- being aware that some pupils will take longer to complete tasks, including homework;
- taking into account the higher levels of concentration and physical exertion required by some pupils (particularly for prolonged practical work) that will lead to increased fatigue for pupils who may already have reduced stamina;
- being aware of the extra effort required by some pupils to follow oral work, whether through use of residual hearing, lip reading or signed support and of the tiredness and limited concentration which is likely to ensue;
- ensuring all pupils are included, and can participate safely, in school trips and off-site visits.

These, and other more specific strategies, are placed in the context of supporting particular individuals described in the case studies in Chapter 8.

The success of design and technology

In general, teachers of pupils with SEND welcome the opportunities offered by National Curriculum D&T to consolidate the learning of skills which are not so readily accessible in other subjects:

> D&T makes a crucial contribution to the development of pupils' practical and thinking skills, and their ability to manufacture products, using a range of materials, tools equipment and techniques. It is this essentially

practical aspect that makes D&T an attractive and valuable learning experience and environment for pupils of all abilities. D&T enables pupils to draw upon knowledge and understanding from across the curriculum as well as from D&T, and to apply this in a very practical way. For this reason D&T is extremely valuable to the learning and progress of pupils with SEN.

Curriculum Council for Wales (1993)
Design & Technology: One in five, Cardiff: CCW

Pupils with SEND make better progress in D&T than in most other subjects. Pupils enjoy the practical application and can see the results of their efforts easily. The projects and lessons lend themselves to effective differentiation:

The needs of all pupils in D&T were met well in the highest performing primary and secondary schools. Pupils with special educational needs and/or learning disabilities and lower-attaining students made good progress as a result of the good individual support they received. Most pupils in all of the schools visited enjoyed designing and making products, solving problems and seeing their ideas taking shape.

(Ofsted, *Meeting technological challenges?*
Design and technology in schools, 2007–10)

2 Starting points

Different types of SEND in D&T

This section describes the main characteristics of particular types of learning difficulties with practical ideas for how a design and technology teacher can help, and contacts for further information. Some of the tips are based on good secondary practice while others encourage teachers to try new, or less familiar, approaches.

Take care in using labels when talking with parents, pupils or other professionals. Unless a pupil has a firm diagnosis, and parents and pupil understand the implications of that diagnosis, it is more appropriate to describe the features of the special educational need rather than use the label; for example a teacher might describe a pupil's spelling difficulties, but not use the term 'dyslexic'.

Consider the pupil with SEN as an individual within your school and subject environment. The strategies in this chapter may help teachers to adapt that environment to meet the needs of individual pupils within the subject context. For example, rather than saying, 'He can't read the worksheet', recognise that the worksheet is too difficult for the pupil, and adapt the work accordingly.

There is a continuum of need within each of the special educational needs listed here. Some pupils will be affected more than others, and show fewer or more of the characteristics described. This continuum of need will also impact on the design and technology teacher's planning and allocation of support staff.

Communication and interaction

Speech, language and communication needs (SLCN)

Pupils with SLCN have problems understanding what others say and/or making others understand what they say. Their development of speech and language skills may be significantly delayed. Speech and language difficulties are common in young children, but most problems are resolved during the primary years. Problems that persist beyond the transfer to secondary school will be more severe. Any problem affecting speech, language and communication will have a significant effect on a pupil's self-esteem, and personal and social relationships. The development of their literacy skills is also likely to be affected. Even where pupils learn to decode, they may not understand what they have read. Sign language gives pupils an additional method of communication. Pupils with speech, language and communication difficulties cover the whole range of academic abilities.

Main characteristics

- **Speech difficulties**
  Difficulties with expressive language may involve problems in articulation and the production of speech sounds, or in coordinating the muscles that control speech. Pupils may have a stammer or some other form of dysfluency.

- **Language/communication difficulties**
  Receptive language impairments lead to difficulty in understanding other people. Pupils may use words incorrectly with inappropriate grammatical patterns, have a reduced vocabulary, or find it hard to recall words and express ideas. Some pupils will also have difficulty using and understanding eye contact, facial expression, gesture and body language.

How can the design and technology teacher help?

- Talk to parents, the speech therapist and the pupil.
- Learn the most common signs for your subject.
- Use visual supports: objects, pictures and symbols.
- Use the pupil's name when addressing them.
- Give one instruction at a time, using short, simple sentences.
- Give pupils time to respond before repeating a question.
- Make sure pupils understand what they have to do before starting a task.
- Provide alternatives for communication, for example: a scribe to record designs and views about products.
- Design directly with the materials, giving pupils choice.

- Pair pupils with a work/subject buddy.
- Give pupils access to a computer or other IT equipment appropriate to the subject.
- Give pupils written homework instructions.

Further information

ICAN, 31 Angel Gate (Gate 5), Goswell Road. London, EC1V 2PT.
Website: www.ican.org.uk
AFASIC, 1st Floor, 20 Bowling Green Lane, London EC1R 0BD.
Helpline: 0300 666 9410
Tel: 020 7490 9410
Fax: 020 7251 2834
Email: info@afasic.org.uk
Website: www.afasic.org.uk

Autistic Spectrum Disorders (ASD)

The term 'Autistic Spectrum Disorders' is used for a range of disorders affecting the development of social interaction, social communication, social imagination and flexibility of thought. This is known as the 'Triad of Impairments'. Pupils with ASD cover the full range of ability and the severity of the impairment varies widely. Some pupils also have learning disabilities or other difficulties. Four times as many boys as girls are diagnosed with an ASD.

Main characteristics

- **Social interaction**

  Pupils with an ASD find it difficult to understand social behaviour and this affects their ability to interact with children and adults. They do not always understand social contexts. They may experience high levels of stress and anxiety in settings that do not meet their needs or when routines are changed. This can lead to inappropriate behaviour.

- **Social communication**

  Understanding and use of non-verbal and verbal communication is impaired. Pupils with an ASD have difficulty understanding the communication of others and in developing effective communication themselves. They have a literal understanding of language. Many are delayed in learning to speak, and some never develop speech at all.

- **Social imagination and flexibility of thought**

  Pupils with an ASD have difficulty in thinking and behaving flexibly which may result in restricted, obsessional, or repetitive activities. They are often more interested in objects than people, and have intense interests in one particular area such as trains or vacuum cleaners. Pupils work best when they have a routine. Unexpected changes in those routines will cause distress. Some pupils with Autistic Spectrum Disorders have a different perception of sound, sight, smell, touch, and taste, and this can affect their response to these sensations.

  Pupils with an ASD may find D&T particularly difficult because it involves social interaction, social communication and flexibility of thought to complete every design and make projects in the practical classroom. However, each case is quite unique and some pupils with an ASD have been known to excel and be gifted in specific areas, for example designing, and observing real products. The teacher must balance allowing the pupil to develop their unique talents and ensuring sufficient development across all areas.

How can the design and technology teacher help?

- Liaise with parents, as they will have many useful strategies.
- Provide visual supports in class: objects, pictures, etc.
- Give a symbolic or written timetable for each day.
- Give advance warning of any changes to usual routines.
- Warn pupils when they will need to be flexible and anticipate some distress, for example a food product may take between 10–15 minutes to cook and the pupils may become distressed if it is not ready exactly on time.
- Provide pupils with their own practical workspace if possible, and allow them access to computers for designing.
- Avoid using too much eye contact as it can cause distress.
- Give individual instructions using the pupil's name, e.g. 'Paul, bring me your design folder'.
- Individual projects, related to pupils' own interests, may be more successful than those set outside their experience requiring them to work as a team.
- Structured projects that guarantee success will build confidence.
- Avoid using metaphor, idiom or sarcasm – say what you mean in simple language.

Further information

The National Autistic Society, 393 City Road, London ECIV 1NG.
Tel: 0808 800 4104
Website: www.autism.org.uk

Asperger's Syndrome

Asperger's Syndrome is a type of autism. People with Asperger's Syndrome have average to high intelligence but share the same Triad of Impairments. They often want to make friends, but do not understand the complex rules of social interaction. They have impaired fine and gross motor skills, with writing being a particular problem. Boys are more likely to be affected – with the ratio being 10:1 boys to girls. Since they appear 'odd' and naïve, these pupils are particularly vulnerable to bullying.

Main characteristics

- **Social interaction**
  Pupils with Asperger's Syndrome want friends but have not developed the strategies necessary for making and sustaining friendships. They find it very difficult to learn social norms and to pick up on social cues. Highly social situations, such as lessons, can cause great anxiety.

- **Social communication**
  Pupils have appropriate spoken language but tend to sound formal and pedantic, using little expression and with an unusual tone of voice. They have difficulty using and understanding non-verbal language such as facial expression, gesture, body language and eye contact. They have a literal understanding of language and do not grasp implied meanings.

- **Social imagination**
  Pupils with Asperger's Syndrome need structured environments, and to have routines that they understand and can anticipate. They excel at learning facts and figures, but have difficulty understanding abstract concepts and in generalising information and skills. They often have all-consuming special interests.

How can the design and technology teacher help?

- Liaise closely with parents, especially over homework.
- Create as calm a classroom environment as possible, particularly during practical lessons.
- Allow the pupil to sit in the same place for each lesson.
- Set up a work buddy system for your lessons.
- Provide additional visual cues in class, such as a step-by-step flow diagram with pictures of how to make the product.
- Give the pupil time to process questions and respond.
- Make sure pupils understand what you expect of them.
- Offer alternatives to handwriting for recording work.

- Use visual timetables and task activity lists.
- Prepare for changes to routines well in advance.
- Give written homework instructions.
- Have your own class rules and apply them consistently.

Further information

The National Autistic Society, 393 City Road, London ECIV 1NG.
Tel: 0808 800 4104
Website: www.autism.org.uk

Semantic Pragmatic Disorder (SPD)

Semantic Pragmatic Disorder is a communication disorder which falls within the autistic spectrum. 'Semantic' refers to the meanings of words and phrases and 'pragmatic' refers to the use of language in a social context. Pupils with this disorder have difficulties understanding the meaning of what people say and in using language to communicate effectively. Pupils with SPD find it difficult to extract the central meaning – saliency – of situations.

Main characteristics

- delayed language development
- fluent speech but may sound stilted or over-formal
- may repeat phrases out of context from videos or adult conversations
- difficulty understanding abstract concepts
- limited or inappropriate use of eye contact, facial expression or gesture
- problems with motor skills

How can the design and technology teacher help?

- Get the pupil to sit or work at the front of the room, to avoid distractions.
- Use visual supports such as objects, pictures or symbols.
- Pair pupils with a work/subject buddy.
- Create a calm working environment with clear rules.
- Be specific and unambiguous when giving instructions.
- Make sure instructions are understood, especially when using subject-specific vocabulary that can have another meaning in a different context, for example 'fold', 'shape', 'saw', 'bend', 'seam', 'flour'.

Further information

AFASIC, 1st Floor, 20 Bowling Green Lane, London EC1R 0BD.
Helpline: 0300 666 9410
Tel: 020 7490 9410
Fax: 020 7251 2834
Email: info@afasic.org.uk
Website: www.afasic.org.uk

Tourette's Syndrome (TS)

Tourette's Syndrome is a neurological disorder characterised by 'tics'; involuntary, rapid or sudden movements or sounds that are frequently repeated. There is a wide range of severity of the condition with some people having no need to seek medical help whilst others have a socially disabling condition. The tics can be suppressed for a short time, but will be more noticeable when the pupil is anxious or excited.

Main characteristics

Physical tics

These range from simple blinking or nodding through more complex movements and conditions such as echopraxia (imitating actions seen) or copropraxia (repeatedly making obscene gestures).

Vocal tics

Vocal tics may be as simple as throat clearing or coughing, but can progress to be as extreme as echolalia (the repetition of what was last heard) or coprolalia (the repetition of obscene words).

TS itself causes no behavioural or educational problems but pupils may also present with associated disorders such as attention deficit hyperactivity disorder (ADHD) or obsessive compulsive disorder (OCD) may be present.

How can the design and technology teacher help?

- Establish a good rapport with the pupil.
- Talk to the parents about TS.
- Agree an 'escape route' signal should the tics become disruptive for the rest of the class.
- Allow the pupil to work at the back of the room to prevent other pupils staring.
- Give the pupil access to a computer to reduce the need for handwriting.
- Design directly with the materials.
- Make sure the pupil is not teased or bullied.
- Be alert for signs of anxiety or depression.

Further information

Tourette's Action
Tel: 0300 777 8427
Website: www.tourettes-action.org.uk

Cognition and learning

Specific learning difficulties (SpLD)

The term 'specific learning difficulties' covers dyslexia, dyscalculia and dyspraxia.

Dyslexia

The term 'dyslexia' is used to describe a learning difficulty associated with words that can affect a pupil's ability to read, write and/or spell. Research has shown that there is no one definitive definition of dyslexia or one identified cause, and it has a wide range of symptoms. Although found across a whole range of ability levels, the idea that dyslexia presents as a difficulty between expected outcomes and performance is widely held.

Main characteristics

Cognition and learning

The pupil may frequently lose their place while reading, make a lot of errors even with high frequency words, have difficulty reading names, blending sounds and segmenting words. Reading requires a great deal of effort and concentration.

Written work may seem messy, with jumbled words (tired/tried) and crossing out. Similarly shaped letters may be confused, such as b/d/p/q, m/w and n/u. Spelling difficulties often persist into adult life and these pupils become reluctant writers.

How can the design and technology teacher help?

- Be aware of the pupil's strengths and specific areas of difficulty.
- Keep instruction sheets simple – use step-by-step flow diagrams with visual clues, such as pictures of each process and one action per step.
- Provide key word sheets and writing frames to support designing, planning, making and evaluating.
- Teach and allow the use of word-processers, spell checkers and computer-aided learning packages.
- Provide word lists and photocopies, rather than directing pupils to copy from the board.
- Consider alternatives to writing, e.g. pictures, plans, flow charts and mind maps.
- Allow pupils extra time for tasks, including assessments and examinations.

Further information

The British Dyslexia Association
Tel: 0333 405 4567
Website: www.bdadyslexia.org.uk

Dyslexia Action
Tel: 03003038840
Website: www.dyslexiaaction.org.uk

Dyslexia Association
Website: www.dyslexia.uk.net
Tel: 0115 924 6888

Dyscalculia

The term 'dyscalculia' is used to describe difficulty with mathematics. This might be a marked discrepancy between the pupil's developmental level and general ability on measures of specific maths ability, or a total inability to abstract or consider concepts and numbers.

Main characteristics

- The pupil may have difficulty counting by rote, writing or reading numbers, may miss out or reverse numbers, have difficulty with mental maths, and be unable to remember concepts, rules and formulae.
- In maths-based concepts, the pupil may have difficulty with money, telling the time, giving/following directions, using right and left and sequencing events. They may lose track of turntaking, e.g. in team games or sharing equipment.

How can the design and technology teacher help?

- Provide number/word/rule/formulae lists and photocopies, rather than directing pupils to copy from the board.
- Use weighing scales and measuring equipment that is clear and easy to read. Use simple measuring boards and cards so that the pupil does not have to deal with all markings on rulers.
- Make use of ICT and teach the use of calculators.
- Encourage the use of rough paper for working out.
- Plan the layout of work, with it well spaced on the page.
- Provide practical objects that are appropriate to the pupil's age to aid learning.
- Allow extra time for tasks including assessments and examinations.

Further information

Dyscalculia
Website: www.dyscalculia.co.uk

British Dyslexia Association
Tel: 0333 405 4555
Website: www.bdadyslexia.org.uk

Dyscalculia Centre
Website: www.dyscalculia.me.uk
Tel: 01604 880 927

Dyspraxia

Dyspraxia is a common developmental disorder that affects fine and gross motor coordination and may also affect speech. The pattern of coordination difficulties will vary from person to person and will affect participation and functioning in everyday life as well as in school.

Main characteristics

- difficulty in co-ordinating movements, may appear awkward and clumsy
- difficulty with handwriting and drawing, throwing and catching
- difficulty following sequential events, e.g. multiple instructions
- may misinterpret situations, take things literally
- limited social skills, resulting in frustration and irritability
- possible articulation difficulties

How can the design and technology teacher help?

- Be sensitive to the pupil's limitations when co-ordination and dexterity are required in designing and making, and plan tasks to enable success.
- Ask the pupil questions to check their understanding of instructions/tasks.
- Help the pupils to sequence their activities with a step-by-step flow chart.
- Consider the best position for a pupil to watch a demonstration; think about having the pupil on the same side as the teacher.
- Check the pupil's seating position to encourage good presentation (both feet resting on the floor, desk at elbow height and ideally with a sloping surface to work on).
- Use grid paper and grid card to help with drawing and modelling.
- Use ready-made templates and guides to speed up work and help with accuracy.
- A light box can help pupils to trace, copy and draw.
- Consider using modified handles and stands to support equipment.

Further information

Dyspraxia Foundation, 8 West Alley, Hitchin, Herts, SG5 1EG.
Tel: 01462 455 016
Website: www.dyspraxiafoundation.org.uk

Moderate learning difficulties (MLD)

The term 'moderate learning difficulties' is used to describe pupils who find it extremely difficult to achieve expected levels of attainment across the curriculum, even with a differentiated and flexible approach. These pupils do not find learning easy and can suffer from low self-esteem and sometimes exhibit unacceptable behaviour as a way of avoiding failure.

Main characteristics

- difficulties with reading, writing and comprehension
- problems understanding and retaining basic mathematical skills and concepts
- immature social and emotional skills
- limited vocabulary and communication skills
- short attention span
- underdeveloped co-ordination skills
- lack of logical reasoning
- inability to transfer and apply skills to different situations
- difficulty remembering what has been taught previously
- difficulty with peronsal organisation such as following a timetable, remembering books and equipment

How can the design and technology teacher help?

- Discover the pupil's strengths, weaknesses and attainment levels.
- Establish a routine within the lesson.
- Keep tasks short and varied.
- Keep listening tasks short or interspersed with activities.
- Provide word lists, writing frames and shortened text for reading.
- Try alternative methods of recording information, e.g. drawings, charts, labelling, diagrams, use of ICT.
- Check previously gained knowledge and build on it (pupils will vary from one another).
- Repeat information and instructions in different ways.
- Show the pupil what to do, or what the expected outcome is, and demonstrate practical skills or show examples of completed work.
- Use practical, concrete, visual examples to illustrate explanations.
- Design directly with the materials.
- Lead and feed ideas, showing finished examples so that the pupil sees the point of designing and making.
- If the ideas developed are impractical and cannot be made by the pupil, explain to them why they might not be able to realise their ideas, but highlight as many positive parts of the ideas as possible.

- If the pupil has a limited range of ideas and experience to draw on for design concepts, provide a range of age-appropriate stimuli and some possible ideas for the pupils to develop.
- Use jigs, templates, patterns, pre-cut or pre-made parts if co-ordination and accuracy are issues.
- Question the pupil to check they have grasped a concept or can follow instructions.
- Make sure the pupil always has something to do.
- Use lots of praise, instant rewards, catch them trying hard.

Further information

MENCAP
Tel: 0808 808 1111
Website: www.mencap.org.uk

Severe learning difficulties (SLD)

This term covers a wide and varied group of pupils who have significant intellectual or cognitive impairments. Many have communication difficulties and/or sensory impairments, in addition to more general cognitive impairments. They may also have difficulties in mobility, coordination and perception. Some pupils may use signs and symbols to support their communication and understanding.

How can the design and technology teacher help?

- Liaise with parents (perhaps through the SENCO initially).
- Arrange a work/subject buddy.
- Use visual supports: objects, pictures, symbols.
- Learn some signs relevant to the subject (D&T in this case).
- Allow the pupil time to process information and formulate responses.
- Set differentiated tasks linked to the work of the rest of the class.
- Lead and feed ideas showing finished examples so that the pupil sees the point of designing and making.
- If the pupil has a limited range of ideas and experience to draw on for design ideas, provide a range of age-appropriate stimuli and some possible ideas for the pupils to develop.
- Design directly with the materials.
- Use jigs, templates, patterns, pre-cut or pre-made parts if coordination and accuracy are issues.
- Set achievable targets for each lesson or module of work.
- Accept different recording methods: drawings, audio or video recordings, photographs, etc.
- Give access to computers where appropriate.
- Give a series of short, varied activities within each lesson.

Further information

MENCAP
Tel: 0808 808 1111
Website: www.mencap.org.uk

Down's Syndrome (DS)

Down's Syndrome is the most common identifiable cause of learning disability. This is a genetic condition caused by the presence of an extra chromosome 21. People with DS have varying degrees of learning difficulties ranging from mild to severe. They have a specific learning profile with characteristic strengths and weaknesses. All share certain physical characteristics but will also inherit family traits in physical features and personality. They may have additional sight, hearing, respiratory, and heart problems.

Main characteristics

- delayed motor skills
- taking longer to learn and consolidate new skills
- limited concentration
- difficulties with generalisation, thinking and reasoning
- sequencing difficulties
- stronger visual than aural skills
- better social than academic skills

How can the design and technology teacher help?

- Sit or position the pupil in a practical room where they can best see and hear.
- Speak directly to the pupil, reinforce speech with facial expression, pictures, objects.
- Use simple, familiar language in short sentences.
- Check instructions have been understood.
- Give the pupil time to process information and formulate a response.
- Give the pupil plenty of time to practise basic practical skills and reinforce ones learned earlier.
- Practise sequencing with simple stages of a process – muddled up and placed in order with the pupil.
- Break lessons up into a series of shorter, varied and achievable tasks.
- Accept other ways of recording: drawings, tape/video recordings, symbols, etc.
- Set differentiated tasks linked to the work of the rest of the class.
- Lead and feed ideas showing finished examples so that the pupil sees the point of designing and making.
- If the ideas developed are impractical and cannot be made by the pupil, explain to them why they might not be able to realise their ideas, but highlight as many positive parts of the ideas as possible.
- Provide age-appropriate resources and activities.
- If the pupil has a limited range of ideas and experience to draw on for design ideas provide a range of age-appropriate stimuli and some possible ideas for the pupils to develop.

- Use jigs, templates, patterns, pre-cut or pre-made parts if coordination and accuracy are issues.
- Allow the pupil to work in groups with more able peers to give good behaviour models.
- Provide a work buddy.
- Expect the pupil to work unsupported for part of each lesson.

Further information

The Down's Syndrome Association
Tel: 0333 1212 300
Website: www.downs-syndrome.org.uk

Fragile X Syndrome

Fragile X Syndrome is caused by a malformation of the X chromosome and is the most common form of inherited learning disability. This intellectual disability varies widely, with up to a third of those diagnosed having learning problems ranging from moderate to severe. More boys than girls are affected but both may be carriers. Strengths associated with Fragile X include good imitation skills, visual learning and long-term memories. These traits can help their learning in a practical subject such as D&T.

Main characteristics

- delayed and disordered speech and language development
- difficulties with the social use of language
- articulation and/or fluency difficulties
- verbal skills better developed than reasoning skills
- repetitive or obsessive behaviour, such as hand-flapping, chewing, etc.
- clumsiness and fine motor coordination problems
- attention deficit and hyperactivity
- may easily become anxious or overwhelmed in busy environments

How can the design and technology teacher help?

- Liaise with parents (perhaps through the SENCO initially).
- Make sure the pupil knows what is planned for each lesson – provide visual timetables, work schedules or written lists.
- Ensure the pupil sits or works in practical lessons at the front of the class, in the same seat for all lessons.
- Arrange a work/subject buddy.
- Where possible, keep to routines and give prior warning of all changes.
- Make instructions clear and simple to understand.
- Use visual supports: objects, pictures, symbols, etc.
- Provide checklists and plenty of discussions to help with designing and evaluating products. Help the pupil to evaluate and discuss some favoured products.
- Design directly with the materials.
- Allow the pupil to use a computer to record and access information.
- Give lots of praise and positive feedback.

Further information

Fragile X Society, Rood End House, 6 Stortford Road, Dunmow, CM6 1DA.
Tel: 01371 875100
Website: www.fragilex.org.uk

Profound and multiple learning difficulties (PMLD)

Pupils with profound and multiple learning difficulties have complex learning needs. In addition to very severe learning difficulties, pupils have other significant difficulties, such as physical disabilities, sensory impairments or severe medical conditions. Pupils with PMLD require a high level of adult support, both for their learning needs and for personal care.

Pupils with PMLD are able to access the curriculum largely through sensory experiences and stimulation. Some pupils communicate by gesture, eye pointing or symbols, others by very simple language. Some pupils will make small steps of progress, no progress or may even regress because of associated medical conditions. For this group, experiences are as important as academic attainment.

How can the design and technology teacher help?

- Liaise with parents and support assistants.
- Consider the classroom layout and equipment provided.
- Identify possible sensory experiences in your lessons.
- Use additional sensory supports: objects, pictures, fragrances, music, movements, food, etc.
- Lead and feed ideas, showing finished examples so that the pupil sees the purpose of the task.
- Provide a range of age-appropriate stimuli and some possible ideas for the pupils to develop.
- Use adapted equipment.
- Use jigs, templates, patterns, pre-cut or pre-made parts.
- Design directly with the materials, giving choices from a range appropriate to the pupil.
- Take photographs to record experiences and responses.
- Set up a work/subject buddy-rota for the class.
- Identify times when the pupil can work with groups.

Further information

MENCAP   Tel: 0808 808 1111   Website: www.mencap.org.uk

Social, emotional and mental health difficulties

This area includes social, emotional and mental health difficulties and attention deficit disorder with or without hyperactivity. These difficulties can be seen across the whole ability range and have a continuum of severity. Pupils with special educational needs in this category are those who have persistent difficulties despite an effective school behaviour policy and a personal and social curriculum.

Attention deficit disorders (with or without hyperactivity) (ADD/ADHD)

Attention deficit hyperactivity disorder is a term used to describe children who exhibit over-active behaviour and impulsivity and who have difficulty in paying attention. It is caused by a form of brain dysfunction of a genetic nature. ADHD can sometimes be controlled effectively by medication and can be prevalent in children of all levels of ability.

Main characteristics

- difficulty in following instructions and completing tasks
- easily distracted by noise, movement of others, objects attracting attention
- difficulty in listening
- restlessness, an inability to keep still, cause fidgeting
- difficulty controlling behaviour; interferes with other pupils' work
- can't stop talking, interrupts others, calls out
- may move around the room at inappropriate times
- difficulty in waiting or taking turns
- impulsivity – acting without thinking about the consequences

How can the design and technology teacher help?

- Make eye contact and use the pupil's name when speaking to him.
- Keep instructions simple – the one sentence rule.
- Provide clear routines and rules, rehearse them regularly, for example getting ready for practical work, clearing away.
- Sit the pupil away from obvious distractions, e.g. windows, the computer, machinery.
- In busy situations and practical lessons, direct the pupil by name to visual or practical objects.
- Encourage the pupil to repeat instructions back to you before starting a task.
- Tell the pupil when to begin a task clearly.
- Give two choices – avoid the option of the pupil saying 'no': e.g. ' Do you want to use the blue or red acrylic for your holder?'

- Give advance warning when something is about to happen. Signal the change or finish with a time, e.g. 'In two minutes I need you (pupil name) to . . .'.
- Give specific praise – catch him being good, give attention for positive behaviour.
- Give the pupil responsibilities so that others can see him in a positive light and he develops a positive self-image.

Further information

Attention Deficit Disorder Association
Tel: 0800 939 1019
Website: www.add.org

Mental health difficulties (anxiety, depression, self harming, substance abuse or eating disorders)

One in ten children and young people aged 5–16 has a clinically diagnosed mental health disorder and around one in seven has less severe problems. The World Health Organization (WHO) defines mental health as not simply the absence of disorder but 'a state of well-being in which every individual realises his or her own potential, can cope with the normal stresses of life, can work productively and fruitfully, and is able to make a contribution to her or his community'. (*WHO: Mental health: a state of wellbeing.* Geneva: World Health Organization, 2011.) Mental health problems in children and young people cause distress and can have wide-ranging effects, including impacts on educational attainment and social relationships, as well as affecting life chances and physical health. Mental health problems in children and young people can be long-lasting. It is known that fifty per cent of mental illness in adult life (excluding dementia) starts before age fifteen and seventy-five per cent by age eighteen. In addition, there are well-identified increased physical health problems associated with mental health difficulties.

Main characteristics

- inattentive, poor concentration and lack of interest in school/school work
- easily frustrated, anxious about changes
- difficulty working in groups
- difficulty working independently, constantly seeking help
- confrontational – verbally and physically aggressive towards pupils and/ or adults
- destructive – may damage their own or others' property
- moody – may appear withdrawn, distressed, unhappy, sulky; may self-harm
- lacking confidence, acts extremely frightened, lacking self-esteem
- difficulty communicating with other pupils and adults
- difficulty accepting praise from peers or adults

How can the design and technology teacher help?

- Monitor the ability level of the pupil and adapt the level of work to this. Support the pupil's basic skills development, such as reading and writing with writing frames or recording pro-formas.
- Consider the pupil's strengths and interests and make use of them in a design and make assignment.
- Choose a design and make assignment where pupils are guaranteed success and make a high-quality product using CAD/CAM.
- In cases where a pupil only wants to 'make', choose a task that will only work if some design work is carried out. Show importance of design in the pupil's favourite products and things of interest to them.

- Make the pupil aware of your expectations regarding their work and behavior in advance.
- Talk to the pupil to learn more about them; try to ascertain their likes and dislikes.
- Position the pupil close to you during practical work, or close to supportive peers.
- Create a reward system for achieving targets, for example have a list of ten tasks that you expect them to complete for the project and tick them off during the lesson.
- Use demonstrations, model thinking and step-by-step planners to show pupils what to do next and encourage independence.
- Where pupils produce few/stereotypic ideas because they do not want to risk failure, provide plenty of ideas, alternatives and stimuli.
- Focus your comments on the behaviour, not on the pupil, and offer an alternative way of behaving when correcting the pupil.
- Use positive language and verbal praise whenever possible.
- If a pupil destroys their work, or struggles when they make mistakes, highlight the mistakes and experiences of well-known designers. Show how mistakes can be corrected in order to remove the pupil's fear of making them.
- If pupils fear judging their work against high achievers when evaluating their own products, make sure it is clear that they should be judging against the design specification. Develop supportive peer evaluation.
- *Tell* the pupil what you want them to do: 'I need you to . . .', 'I want you to . . .', rather than asking them. Telling them avoids confrontation and negates the possibility that there is room for negotiation.
- Give the pupil a choice between two options.
- Be consistent with pupils and stick to what you say.
- Give the pupil (realistic and achievable) responsibilities to increase their self-esteem and confidence.
- Plan a 'time out' system. Ask a colleague for help with this if you have not done it before.

Further information

SEBDA   Tel: 01233 622958   Website: www.sebda.org

Sensory and/or physical needs

Physical disability (PD)

There is a wide range of physical disabilities, and pupils with PD span all academic abilities. Some pupils are able to access the curriculum and learn effectively without additional educational provision. For other pupils, the impact on their education may be significant, and the school will need to make adjustments to enable them to access the curriculum.

Some pupils with a physical disability have associated medical conditions that may have an impact on their mobility. These include cerebral palsy, heart disease, spina bifida and hydrocephalus, and muscular dystrophy. Pupils with physical disabilities may also have sensory impairments, neurological problems or learning difficulties. They may use a wheelchair and/or additional mobility aids. Some pupils will be mobile but may have significant fine motor difficulties that require support or specialist resources. Others may need augmentative or alternative communication aids.

Pupils with a physical disability may need to miss lessons to attend physiotherapy or medical appointments. They are also likely to become very tired as they expend greater effort to complete everyday tasks. Teachers need to be flexible and sensitive to individual pupil needs.

How can the design and technology teacher help?

- Get to know pupils (and parents) so that they can help you make the appropriate adjustments.
- Maintain high expectations.
- Consider the classroom layout and provision of equipment: plan for provision and access.
- Choose projects carefully, incorporate pupil interests and experiences where possible.
- Provide extra time to complete complex tasks and, if feasible, set less complex, shorter tasks for pupils that still meet the requirements of the curriculum.
- Allow the pupil to leave lessons a few minutes early to avoid busy corridors and give them time to get to next lesson.
- Get the pupil to collect all materials and tools together before they start a task so that they do not need to fetch things repeatedly.
- Set homework earlier in (or at the beginning of) the lesson so instructions are not missed.
- Speak directly to the pupil, rather than through a teaching assistant.
- Allow a helper to assist under the pupil's instruction, for example: a scribe to record designs, or a helper for making items.

- Use jigs, templates, pre-made parts, CAD/CAM to facilitate construction and support the accuracy and quality of outcome.
- Allow pupils to make their own decisions.
- Ensure access to appropriate IT equipment for the duration of the lesson – and see that it is used!
- Allow students to make use alternative methods to record their work, for example: recording a video, instead of writing.
- Plan ahead to cover work missed through medical or physiotherapy appointments.
- Be sensitive to fatigue, especially at the end of the school day.
- Reward all forms of progress observed in the pupil.

Cerebral palsy (CP)

Cerebral palsy is a persistent disorder of movement and posture. It is typically caused by an injury to the brain before or during birth, or in early childhood. Problems vary from slight clumsiness to more severe lack of control of movements. Pupils with CP may also have learning difficulties. They may use a wheelchair or other mobility aid.

Main characteristics

There are three main forms of cerebral palsy:

- *spasticity* – disordered control of movement associated with stiffened muscles;
- *athetosis* – frequent involuntary movements;
- *ataxia* – an unsteady gait with balance difficulties and poor spatial awareness;

Pupils with CP may also have communication difficulties.

How can the design and technology teacher help?

- Gather information from parents, the physiotherapist – and the pupil – about how it affects them specifically.
- Consider the classroom layout and equipment to be used, for example access to workspaces in practical lessons, use of CAD/CAM, adaptions to sewing machines, etc.
- Get the pupil to collect all materials and tools together before they start so that they do not need to fetch things repeatedly.
- Have high academic expectations.
- Use jigs, templates, patterns, pre-cut or pre-made parts if coordination and accuracy are issues.

- Use visual supports: objects, pictures, symbols, etc.
- Arrange a work/subject buddy or helper in the class to work under the direction of the pupil.
- Speak directly to the pupil rather than through a teaching assistant.
- Ensure access to appropriate IT equipment for the subject – and see that it is used.

Further information

Scope
Tel: 0808 800 3333
Website: www.scope.org.uk

Cerebral Palsy Org
Tel: 0800 692 4453
Website: www.cerebralpalsy.org.uk

Hearing impairment (HI)

The term 'hearing impairment' is a generic term used to describe all hearing loss. The main types of loss are monaural, conductive, sensory and mixed loss. The degree of hearing loss is described as mild, moderate, severe or profound. Some children rely on lip reading, others will use hearing aids, and a small proportion will have British Sign Language (BSL) as their primary means of communication.

How can the design and technology teacher help?

- Check the degree of loss the pupil has to see how it will impact your lessons.
- Check the best seating position for the pupil (e.g. away from the hum of cookers, fridges, computers, CAD/CAM machines, with the pupil's good ear towards speaker).
- Check that the pupil can see your face for facial expressions and lip reading.
- Provide a list of vocabulary, context and visual clues, especially for new subjects.
- During class discussion allow only one pupil to speak at a time and indicate where the speaker is.
- Check that any aids and other specialist equipment are working.
- Make sure there is adequate lighting so that the pupil can easily see your face and lips. Do not stand with your back to a window. If you use interactive whiteboards, ensure that the beam does not prevent the pupil from seeing your face.
- Ban small talk amongst pupils.

Further information

Action on Hearing Loss (Also known as RNID)
Tel: 0808 808 0123
Website: www.actiononhearingloss.org.uk

British Deaf Association (BDA)
Tel: 020 7697 4140
Website: www.bda.org.uk

British Association of Teachers of the Deaf (BATOD),
Tel: 0845 643 5181
Website: www.batod.org.uk

Visual impairment (VI)

Visual impairment refers to a range of difficulties, including those pupils with monocular vision (vision in one eye), those who are partially sighted and those who are blind. Pupils with visual impairment cover the whole ability range and some pupils may have other special educational needs.

How can the design and technology teacher help?

- Check the optimum position for the pupil, e.g. for a monocular pupil their good eye should be towards the action.
- Always provide the pupil with his own copy of the text.
- Provide enlarged print copies of written text and instructions.
- Check the accessibility of IT systems (enlarged icons, talking text, teach keyboard skills, etc.).
- Do not stand with your back to the window, as this creates a silhouette and makes it harder for the pupil to see you.
- Draw the pupil's attention to displays, which they may not notice.
- Make sure the floor is kept free of clutter.
- Let the pupil know if there is a change to the layout of a space.
- Ask if there is any specialist equipment available that the pupil requires for your subject, such as enlarged print dictionaries, talking weighing scales, Wikki Stix® for designing rather than drawing.
- Consider utilising activities that explore the use of other senses, such as taste testing, in your lessons.

Further information

RNIB Royal National Institute of the Blind
Tel: 0303 123 9999
Website: www.rnib.org.uk

Multi-sensory impairment

Pupils with multi-sensory impairment have a combination of visual and hearing difficulties. They may also have other additional disabilities that make their situation complex. A pupil with these difficulties is likely to require a high level of individual support.

How can the design and technology teacher help?

- The design and technology teacher will need to liaise with support staff to ascertain appropriate provisions for each subject.
- Learn how to use alternative means of communication, as appropriate.
- Be prepared to be flexible and to adapt tasks, targets and assessment procedures.
- Lead and feed ideas showing finished examples, so that the pupil sees the point of designing and making.
- Design directly with the materials.
- If the ideas developed are impractical and cannot be made by the pupil, explain to them why they might not be able to realise their ideas, but highlight as many positive parts of their ideas as possible.
- Provide age-appropriate resources and activities.
- If the pupil has a limited range of ideas and experience to draw on for design concepts, provide a range of age-appropriate stimuli and some possible ideas for the pupils to develop.
- Use jigs, templates, patterns, pre-cut or pre-made parts if coordination and accuracy are issues.

Further information

Sense
Tel: 0300 330 9250 or 020 7520 0999
Website: www.sense.org.uk

Sensory Support Service
Tel: 0117 903 8442
Website: www.sensorysupportservice.org.uk/multi-sensory-impairment

---

3 An inclusive environment for D&T lessons

The value of design and technology for pupils with SEND

Design and technology can be a very popular and valuable subject for pupils with special educational needs because the practical nature of the learning experiences makes it accessible to pupils of all abilities. Pupils draw on their knowledge and understanding from across the curriculum, develop their numeracy, literacy and communication skills, and are required to apply these in practical ways. This helps them to consolidate skills from other lessons and reinforce learning with positive outcomes. Designing and making usable products can give pupils a real sense of achievement as they benefit from seeing clear progress and taking responsibility for their own learning. Their personal engagement with tasks often improves their attention span, patience, persistence and commitment so that, despite their additional needs, pupils can achieve results that are on par with, and occasionally even outshine, their peers. Design and technology offers them the chance to experience achievement at a level that may seldom occur elsewhere in school life.

Design and technology provides particular opportunities for:

- practical learning experiences that promote success and raise pupils' achievement;
- focusing on real scenarios and design problems that are meaningful to pupils;
- equal access to the curriculum through the use of appropriate and differentiated materials that can suit pupils of differing abilities;
- communicating using a variety of methods – avoiding over-reliance on the written word;
- using ICT as a means to pupils fully realising, developing and enhancing their work;
- accessibility and support – D&T helps pupils to reinforce their learning in other subjects, such as maths and science;
- personally motivated tasks – D&T provides pupils with the opportunity for pupils to take ownership of their work and of their own learning;

- working within a flexible range of contexts and topics that can be adapted to suit individual interests and motivations;
- allows pupils to work at a pace and level suitable to them with appropriate teacher support and intervention;
- individually negotiated targets between the teacher and pupil that can be reviewed as required – those pupils who need to work at a slower pace can do so, and pupils who work more quickly can be further challenged to develop their work with activities which extend and enrich their experience.

Planning within D&T

Effective planning in the context of D&T takes account of the different abilities and interests of each pupil and should enable all of them to progress and demonstrate their achievements. The most effective way to secure success is to think positively and create learning opportunities appropriate to each pupil's needs rather than concentrating on identifying difficulties.

Your role as a D&T teacher

Your planning will set out how the curriculum will be adapted to take account of particular learning difficulties.

Adaptations might include:

- making changes to the type of task you provide to particular students;
- providing special support to pupils to carry out the activities successfully;
- modified learning resources that take account of pupils' additional needs.

The examples given in this publication show how units can be adapted to change the type of task or give special support and modified resources to pupils. Considering special educational needs involves teachers in:

- taking account of the difficulties which pupils may encounter in the learning process;
- identifying the impact this may have on pupils' development;
- developing appropriate strategies for supporting pupils with these difficulties.

(The prompt sheets in Appendices 3.1 and 3.2 may help you.)

Table 3.1 Long and medium-term planning

Use the following checklist to help with long and medium-term curriculum planning in your department or faculty:

1. List the learning difficulties of pupils.

2. Indicate any pupils with an education, health, and care plan and the nature of their special provision. Include health and safety implications and medical procedures.

3. Outline any support teacher time available.

4. Outline responsibilities of the support teacher.

5. List any learning support resources available, including local centres or organisations.

6. Indicate how provision for SEND is incorporated into schemes of work and project planning.

7. Identify any specialist SEND CPD/INSET for individual D&T teachers.

8. List any targeted funding for SEND support for D&T.

9. Indicate methods of recording attainments and outline any special provision in relation to exams and tests.

Planning inclusive designing and making

Design and make assignments

Designing and making assignments involve a complex interplay (often called iterative design) between:

- exploring and clarifying the task;
- generating, developing and communicating ideas and proposals;
- testing, evaluating and modifying;
- planning and making.

It is this interplay that makes design and technology particularly demanding for pupils. It is very challenging for a pupil to be developing capability, confidence, independence and decision-making skills if the activity is reduced to a linear teacher-controlled craft activity, where the designing and the outcome is prescribed step by step.

In relation to these processes, pupils will have particular strengths in certain types of work. It is possible to devise activities for some pupils based on their strengths and successes. This may mean centring D&T activities around 'making' and letting other important processes be incorporated through and around making, for example using three-dimensional 'mock-ups' rather than drawings. The interplay of designing and making can, therefore, be a strength as well as a demand for these pupils, where designing is always considered in close relation to making.

Design briefs and projects provide pupils with the chance to put their knowledge and skills to the test to meet challenges that address real needs and wants. Assignments also enable pupils to develop their confidence and ability to apply design ideas and concepts in concrete, practical ways. They will have opportunities to work as individuals or in a team, with opportunities to learn from the work of others. Designing and making assignments or projects should be planned flexibly so that adequate differentiation is possible. All pupils should experience the whole designing and making process from start to finish, but some pupils will complete some aspects in more depth or in a more complex way (enrichment activities), and a few will complete additional tasks (extension activities), where this is appropriate for them.

Pupils often find it easier to work on short, more focused assignments rather than longer, open tasks. Shorter tasks provide small successive elements of success, rewarding and motivating the pupil regularly – the accumulation of success and achievement can be structured to ensure progression:

- Use contexts that pupils are familiar with.
- Adapt or make improvements, or add a new feature to the design of a product rather than 'inventing' a whole new product where their experience is limited.
- Design a product where they are given guidance towards alternative solutions. However, it is important to avoid tokenism – there ought to be an opportunity for real designing and decision taking.
- Manage a project where certain aspects are restricted (for example, the size and shape of a box), but there are still significant opportunities for designing and independent work (for example, designing the puzzle to put in the box, or decorating the outside of the box).
- Join in a project part way through, for example where the research has been completed so that pupils can get into the modelling and making aspects more quickly.

Some examples

- *Salads* – this can require pupils to adapt part of the recipe, such as creating a healthier salad dressing only, rather than changing the whole salad.
- *Container/wallet* – this can require pupils to design a case for their own personal use. It will be a context that is familiar to the pupil so it makes research requirements real but limited.

Focused practical tasks or skills building

Focused practical tasks (FPTs) enable teachers to ensure that pupils practise particular skills and knowledge, consolidating these as they are acquired. They are closely structured and teacher led so that pupils practise or learn a skill or process. FPTs build pupils' confidence and give them ideas for their design work. These 'mini-making' activities are highly motivating as pupils can see the results and progress of their efforts immediately. For example, a teacher may organise a series of activities where the pupil makes samples using a range of fabric decoration techniques before the pupils then consider how to design and make a product that uses one of the fabrics to suit the design brief. Teachers should use short, focused tasks to provide particular outcomes, providing opportunities for pupils to achieve success in one or more D&T processes.

Pupils will find it easier to:

- repeat and reinforce previously learned skills and processes on a regular basis;
- follow instructions independently – due to a simplified set of instructions with clear pictures and diagrams;
- absorb information – by providing small amounts of new information, or a few instructions at one time, the process is broken into smaller, more manageable stages;
- plan their own work – if they have practised this (e.g. putting muddled sets of instructions in the right order) (See the time planner support sheet in Appendix 3.3);
- spell and recognise the names of important pieces of equipment, ingredients and processes, using key word sheets or posters to help them (See Appendix 3.4).

Some examples

- *Salads* – make a range of salads and dressings from set recipes to practise the process and understand how the ingredients behave.
- *Alarms* – use switches and sensors to make circuits from a plan.
- *Containers/wallet* – learning how to make different seams or other joining techniques.
- *Colouring textiles* – learning how to tie-dye and batik.
- *Clocks* – make a clock face using sheet card to practise 3D modelling.

Product evaluation activities

These are activities through which pupils can investigate, disassemble and evaluate products, providing the opportunity to build knowledge, skills and understanding that can be used to inform other D&T activities. Pupils can be encouraged to discuss and evaluate other designers' work against clear criteria.

Pupils will find it easier to

- look at a limited range of products at one time;
- have a mixture of familiar and less familiar products to assess;
- use pro forma worksheets to record their responses;
- discuss, examine and taste products as a group rather than relying on written accounts.

Some examples

Feeding special groups – evaluating some vegetarian products that are available.

Clocks – assessing a selection of clocks made from sheet materials to see how they were constructed.

6Rs – evaluating pairs of products, where one has been made with sustainability in mind.

Supporting designing

For activities focused on the 'making' part of the programme of study, it is assumed that the teacher will often give pupils one-to-one support. Most pupils with special educational needs cope well with the making aspects of work and teachers are effective in providing appropriate guidance. However, particular attention should be paid to facilitating pupils' independence, by utilising materials such as: key words sheets; flow charts to help with the order of making; and simplified, visual instruction sheets which explain a process step-by-step. It is the 'designing' aspect that often proves most challenging to pupils.

Ofsted reports comment every year that 'making' aspects are better taught than 'designing' at Key Stage 3, and the *National Strategy KS3 Foundations Subjects D&T National Dissemination 2004–5* gave teachers further ideas on different designing strategies they could use. Figure 3.1 is an adapted version of the strands of designing, showing how these can be differentiated for abilities and for recording or auditing when they are being taught to your pupils.

As pupils with SEND are likely to require even more support in the designing parts of D&T, particular attention should be paid to planning how pupils can

		Projects				
Colour coding		1	2	3	4	5
Encounter/Aware						
Engagement/Participation						
Involvement/Gaining skill						

Exploring ideas and the task	**Pupils with SEND working towards these:** Observe and explore familiar products and situations, and how things work or are used. Recognise familiar products and explore the different parts they are made from. Operate familiar products, with support and explore how they work Explore familiar products and communicate views about them when prompted. Some may – describe what products are used for and the needs of the people who use them, identify what works well Some may contribute to design and make projects that are linked to their own interests **Some pupils may make further progress and be able to:** Recognise characteristics of familiar products, describe how a product works	
Generating ideas	**Pupils with SEND working towards these:** Respond to a range of sensory experiences They begin to offer responses to making activities, suggesting a colour or shape Explore the qualities of materials by playing and experimenting and trying things out Try out ideas by shaping materials and putting components together Communicate likes and dislikes and demonstrate preferences for products, materials and ingredients **Some pupils may make further progress and be able to:** Generate ideas based on experience of working with materials and components Recognise that their designs have to meet a range of different needs	
Developing and modelling ideas	**Pupils with SEND working towards these:** Make choices about a product or aspects of its design. Observe, explore and experience a range of materials and tools Communicate their designs in a variety of ways, they may draw or model their idea Begin to assemble components provided for the activity. Watch others and copy the actions Explore options within a limited range of materials Some may – Develop their ideas taking into account how their product will be used and who will use them, make products based on the preferences of others **Some pupils may make further progress and be able to:** Use pictures, labelled sketches, models, words to describe their ideas and what they want to do Clarify ideas when asked Add details to designs	
Planning	**Pupils with SEND working towards these:** Respond to options and choices with actions and gestures. They begin to contribute to decisions on what they will do and how. Suggest next steps when planning – they plan by indicating what to do next. Select and use tools and materials from a range suggested by the teacher. Work as an individual on a class project. Work alongside another in the class. Communicate with others in the group. Share equipment Some may – Select and use a variety of tools, equipment and processes **Some pupils may make further progress and be able to:** With help, put their ideas into practice Make plans to achieve aims, think ahead about the order of their work Choose tools, equipment, materials, components and techniques, explain their choices Share ideas with others in the group	
Evaluating	**Pupils with SEND working towards these:** Comment on their work through individual methods of communication. Evaluate against a single concrete criterion, e.g. a photo frame must stand up Some may – Judge the quality of other people's products and identify what works well **Some pupils may make further progress and be able to:** Comment on their own and other people's work and describe how a product works Set their own simple criteria and use these to judge their own product Recognise what they have done well and suggest things they could do better in the future	

Figure 3.1 KS3 Designing strands

Figure 3.2 Modelling with real or
reclaimed materials helps
pupils to design in 3D

Figures 3.3–6 Modelling how things work and how they look with kits, paper, card
and other materials supports pupils' design ideas and makes it more
accessible.

be successful in their initial design work. For example, supporting the recording of ideas quickly; encouraging modelling with materials rather than drawing
ideas; providing stimuli for ideas; accessing and applying simple research
information and presenting and evaluating ideas. (See Appendices 3.5 and 3.6
for example support sheets.)

Diversity in approaches to designing should be valued and teachers should be
aware of adopting too rigid an approach to the way pupils are asked to record
and communicate their design ideas and developments.

Case study: Supporting pupils to generate ideas (puppets)

Debbie Green at John Jamieson School was working with her Year 7 group (8 pupils with a range of abilities and SEND, including medical, physical, emotional and behavioural needs). They were designing puppets to be sold in a local tourist attraction close to the school that had a collection of fish, insects, lizards, butterflies and small monkeys.

She was focusing on teaching two important aspects of designing:

- Using strategies that generate a variety of design ideas quickly as a direct response to design criteria.
- Using a range of strategies to produce, communicate and record initial ideas using a sketchbook and describing.

We had already spent several lessons evaluating similar products, both real examples and ones found on the internet. It just did not work, despite questioning; they produced identical designs or found the task too wide. They were happy to draw an animal, but it was not possible to see how this would become a puppet. I then decided to give them design cards to narrow the task down. The pupils used Widgit software. The cards were made up of two sets – a series of pictures of animals and a series of pictures of different types of puppets (rod, glove, hand, shadow).

This way we now had a game approach to generating ideas. They were in charge and they could choose which cards to pick. We did this as a class first, drawing on a large laminated sheet on the desk; the pupils who could draw participated and the ones who could not were given choices. The task was modelled and then it was their turn individually.

The pupils came up with a variety of designs. Instead of their first attempts (where there was no evidence that the drawing was a puppet), after this scaffolding they produced new designs, for example, clearly showing a snake's body and the rods for the puppet to be attached. They could explain what they were made of and how they worked. Without the cards pupils were restless, with the cards they were focused and on task. The cards generated more detailed and varied designs than in previous lessons. Without the cards most pupils continued to draw what they first thought of, but a slightly different version. One pupil drew five versions of a monkey puppet with split-pin joints – she only began different designs with the cards.

The task was useful because they were able to produce a variety of designs that they could later choose from.

Figure 3.7 Widgit grid

Debbie suggests that during the designing stages teachers should:

- Use design cards displaying choices for the whole class task;
- Give the more able pupils the less structured cards;
- Use word and short sentence banks to enable pupils to label and annotate their drawings;
- Use paper and straws instead of drawing for pupils with a short attention span, as 3D is more fun;
- Use accessible software with symbols for pupils who have difficulty with reading.

Case Study: Becky Wells – Developing design ideas, adapting the 4×4 design strategy for pupils with Autism

Becky Wells, at Sutherland House School, adapted the 4×4 grid that is used in many schools to aid pupils in developing their ideas. Traditionally pupils work in groups of four, each pupil starts by sharing a design idea in the centre of their paper. They then pass the paper to the next person who makes suggestions, comments and sketches to develop or improve the original idea. When pupils work in groups of four, this is done at speed, and the sheet is passed around the group until all four have commented on all four sheets.

Becky adapted this idea to suit her pupils with autism to develop ideas for designing a bag:

- They worked in a linear four-step process.
- Key words and choices were presented at each step.

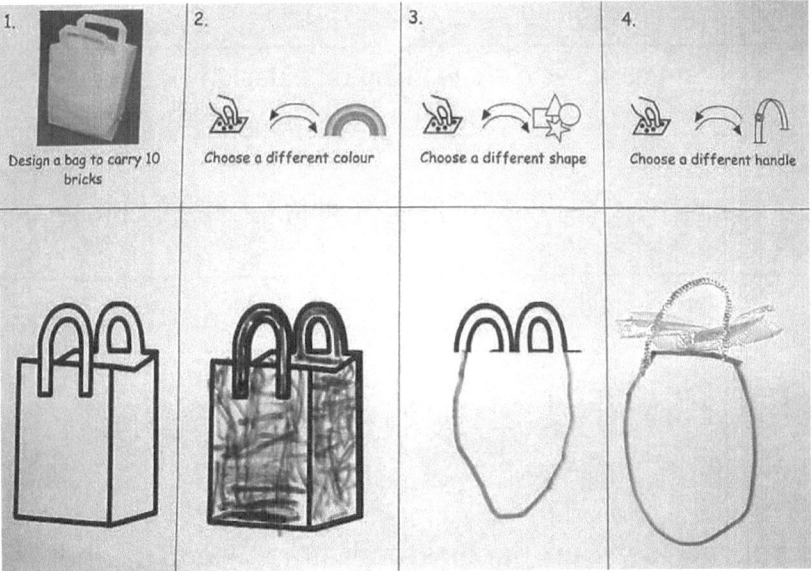

Figure 3.8 Adapted 4 x 4 used to design a bag to carry bricks.

- They practised with a set brief from the teacher – a bag to carry 10 bricks.
- They moved onto developing a bag to carry their favourite toy.

See the website for a slide presentation of this project.

Teachers often report the following difficulties during designing:

1. Clumsiness or difficulties in expressing ideas and in producing drawings.
2. Difficulties in recording thoughts and communicating information.
3. A limited range of ideas and experiences.
4. Difficulty in relating needs and concepts of task to ideas and solutions.
5. Frustration and failure caused by restricted methods of communication imposed – too much written work expected.
6. Pupil unable to connect designing with making.
7. Pupil only wants to make something and not design/draw ideas.
8. Lack of confidence limits ideas: pupil often copies familiar products and plays safe, and quickly becomes disheartened when they see others' innovative design ideas.
9. Ideas may be too narrow or stereotypical.
10. Ideas are unrealistic, impractical and beyond the pupil's making ability.
11. Pupil finds it difficult to sustain momentum on a long project: not interested in detail or accuracy, simply wants to get the project finished quickly.
12. Lacks patience, does not want to revisit a task or improve work, sees little value in testing and evaluating.
13. Pupil is afraid of judging own work against that of their peers or others.
14. Pupil easily loses confidence in work.

Table 3.2 Supporting designing

---

1. Identify a current group who are working on a design and make assignment.

2. What special challenges to the pupils are there during the designing stages of the project?

3. What strategies have you used successfully to help pupils during the designing stages?

Suggested approaches

For example, for a pupil who:

*   finds working ideas on paper too abstract – and cannot relate to the materials or processes;
*   wants to get on and make the product;
*   lacks confidence to present ideas to others.

The following strategies could be used:

*   Give the pupil hands-on activities to encourage more than writing and graphical presentation of ideas.
*   Replace starting points for *drawing ideas* with structured questioning, moving to exploring and trying out ideas with 3D materials. (See Appendix 3.7 for an example of a planned sequence of questions, and Appendix 3.8 for a plenary case study.)
*   Build the pupil's confidence in presenting their ideas by starting as a group rather than individually.
*   Display model thinking for the pupils which explains a designing decision-making process aloud for pupils, for example making a fruit kebab (see Appendix 3.9 for an example of modelling thinking).

Ideas for supporting designing in a number of different projects

*   *Feeding special groups* – the computer can be used for nutritional modelling to compare different recipes before they are made. This saves complex calculations, and results can be shown visually as bar charts.
*   *Container/Wallet* – paper mock-ups or scrap fabric can be used to design the container and work through different ideas.
*   *Colouring textiles* – a template for the product (T-shirt) can be provided (on paper or computer) and pupils can model their ideas with swatches of sample fabrics, paper cuttings and sketches.
*   *Clocks* – thumbnail models can be made in card to explore ideas and

develop them further: bending and shaping card, exploring the limitations of sheet materials.

Strategies for keeping pupils motivated and improving pupil outcomes

Some of the following may seem obvious to experienced teachers. Even so, perhaps you will find some new ideas and that this checklist is useful to you:

- Ensure that pupils' learning objectives are clear and expressed in accessible language.
- Avoid abstract contexts: over a key stage, provide a range of concrete starting points, materials and techniques as some will be of more interest to some pupils than others. Encourage pupils to work on tasks related to their hobbies, interests and strengths.
- Choose projects where pupils are able to produce good quality products, where success is guaranteed and they will be proud of what they have designed and made. This will help pupils' confidence and self-esteem, so that they are able to take risks later with their designing and making.
- There is a fine line between teacher intervention and taking over the pupil's project. The pupil's experience should not solely be one where they are following a designing and making process step-by-step with the teacher doing most of the thinking.
- Ensure that targets match individual's abilities and are challenging but achievable – this will promote self-esteem. The pupil's response to, 'Am I asking too much?' needs to be, 'No, but I am asking a lot!' Provide clear and achievable next steps towards improvement.
- Keep projects short and give pupils regular feedback. Consider the length and complexity of tasks presented: some pupils may be daunted by length, or what they see as a difficult task, and lack the confidence to get started.
- Provide a supportive structure and show pupils a manageable way through the task at hand. Segment the activity into smaller tasks with specific targets – use a tick list or wallchart so that pupils are clear about what they are working towards and where they are in relation to the completion of the project. A project can be broken into smaller steps without being overly prescriptive: instead of broad stages such as 'research', a list of sub-stages could be given such as:

  1. Write five questions for your survey.
  2. Ask your target audience to answer your survey.
  3. Record your results.
  4. Present your results.
  5. Discuss your results with your teacher.

This is motivating where pupils have the opportunity to achieve at least one or two targets in each lesson and keep up the pace. They can be encouraged to tick them off as they achieve each one. Lists of targets can be divided into essential (everyone must do these) and extension (you can chose to do these if you want to, or have time).

- Provide pupils with plenty of stimuli for the project, including a number of design solutions; use classroom visits as a stimulus for design contexts and invite experts into the classroom to work alongside pupils (make sure they are well briefed).

- Use rewards, such as an end-of-project certificate that can be presented in assembly. Use classroom rewards systems such as merit marks, stickers and stamps or personal comments on achievement in folders. (See the website for an example of a project certificate.)

- Reduce the amount of reading, researching and written recording required of pupils. Allow for a greater diversity of learning styles. Give pupils the opportunity to clarify their ideas through discussion rather than relying on writing.

- Pupils who have language difficulties, or experience difficulties in expressing ideas, may be helped by key words (posters, worksheets or labels around the room), flow diagrams and time plans, good use of simple questions, use of worksheets or design prompts with helpful graphics.

- For pupils with literacy difficulties, the use of modelling, role play, video and photographs for recording should be explored to help them to communicate, develop and record their ideas, as well as to interact with a range of communications technologies.

- Difficulties with planning can be aided by pupils being shown good examples of how others have planned (e.g. other pupils or professional designers) so that they can see what good planning entails; or by giving them practice in 'planning retrospectively' before they go on to forward plan. Pupils often find it easier to give an account of what they have done, rather than what they intend to do. Alternatively, they can be given planning sheets to complete (or part-complete) or a list of stages to put in the right order to provide a structure for their thinking. Of course, teacher demonstrations can also help to show children how to order their work and develop spatial relationships.

- For those who only remember a limited number of instructions, and have difficulties listening, avoid giving long introductions and lists of instructions at the beginning of lessons. Ensure that the pupils have an overview of the entire task and then structure the task so that a limited number of instructions are given, and these are carried out before the pupil is given further instructions. Increase the range of sources for the instructions: from worksheets, computer, teacher, wall charts, etc. If on a continuing project, at the end of the lesson, encourage pupils to list what they will do when they come to you next. You can then check this list and add personalised details to it in preparation for the next lesson so the pupils will be able to restart work immediately without waiting for instructions.

- Provide varied examples of successful presentation. Display pupils' work regularly around the school. Encourage the use of word-processing packages and CAD to enhance presentation skills.
- Allow opportunities for some pupils to have extra time to work on their project, perhaps during lunchtime or after school.
- Consider the use of computer-aided manufacture, specially adapted tools and equipment, templates, jigs and patterns and other shortcuts to aid completion of making tasks.

Practical resources to help pupils with SEND

Audio and video recordings

1. Record the pupil's design ideas, comments, plans and evaluations using an audio recorder/player. For some pupils this enables them to express their designing and making in ways that writing or drawing might not.
2. Record the main stages of a process so that the pupils can play it back and use it at their own pace. For example, a teacher may record instructions about how to thread a sewing machine and the pupils can then listen to this with headphones as they set up the machines if they cannot remember how to do it.
3. A 'talking clipboard' and recordable 'idea clouds' (which wipe clean afterwards) are useful for planning and designing ideas.

Using touchscreen/overlay/eye gaze devices

1. Touchscreens, overlay or symbols can be used to trigger a computer. For example, pictures of ingredients and equipment for a design can be chosen in this way, or words for an evaluation can be given.
2. Some pupils find colour-tinted plastic overlay sheets helpful to read and remember certain words. These sheets could be used to highlight particular words or instructions to pupils.
3. Light boxes – The upward shining light can help pupils to trace, copy and draw.

Visual and other labelling to support practical organisation

1. Grid paper with lines at different angles can help pupils to draw in 3D. Grid card is helpful for design modelling, too.
2. Visual reminders and step-by-step explanations of processes (with or without symbols) will be very helpful to all pupils, for example a recipe or method of producing an item developed by the teacher.

3. Digital cameras can be used to take photos of the main stages of a task, with key words added for reference.
4. Large lower-case name labels or Braille labels/symbols and digital photos should be used on equipment cupboards and areas. This will assist all pupils greatly in working independently and organising themselves. They learn the names of the equipment and where they are kept.
5. Avoid storing regularly used items in places that are difficult to reach, such as the back of cupboards, at the top of wall units or in low cupboards.
6. Use carousels and pull-out or pull-down drawers to improve access to places that are difficult to reach.

Organising demonstrations

1. Consider the best position for a pupil to watch a demonstration (good ear towards the demonstrator, for example, if hearing is an issue). Think about having pupils on the same side as the teacher. It is easy to forget that many of the children are seeing the process you are demonstrating from an upside-down position.
2. Sometimes the use of a mirror for left-handed pupils helps them to understand what is being demonstrated, for example the pupil may watch the mirror instead of the teacher when learning how to sew or thread a needle.

Braille and large-print rulers, measuring equipment

1. Specialist equipment suppliers for those with visual impairments will have a range of measuring devices available to schools. Choose weighing scales that are easy to read for all pupils, with clearly marked measures, for example digital scales. Talking weighing scales are also useful.
2. Simple measures (using cups or spoons) can be used for some recipes. Use simple cooker timers that can be set easily to remind pupils of cooking time.
3. Simple measuring boards or cards can be made to help all pupils mark and measure sizes into materials, without having to deal with all the markings on a ruler. For example, a stepped board can be made with standard (or required) sizes for the project – clearly marked with sizes and colour coded. When the pupil uses it, the measure is placed on the material and the length required is marked off. Pupils can quickly see the size and remember the colour more easily.

Practical sessions and safety tips

1. Ensure the pupil can achieve consistent personal organisation, especially in a non-subject specific room. Make sure he or she always works in the same part of the room, thus aiding the development of their mobility skills.

2. To conserve energy, ask the pupil to gather together all the materials and tools they need before they begin the task; or have his or her equipment in a tray or cupboard that is always accessible.

3. Use a perching stool for longer practical sessions.

4. Establish sound safety knowledge. Explain the meaning behind the task to pupils, instead of just telling them. Remember that the role of a support assistant is to promote as much independence as possible, to step in to prevent danger and increase chances of success.

5. Ready-made templates and guides will speed up work and help with accuracy, ensuring a better quality result.

6. Use sticky tape, pads or rubber feet to hold things for pupils who do not have the strength, co-ordination or manual control to hold something still with one hand and perform an operation or process with the other hand.

7. Use a stand or holding jig to clamp the drill in a vertical position, thereby releasing the hands to assist drilling and removing some of the complexity involved in using it. Using a hand drill is quite difficult as there are numerous physical and mental actions involved.

8. A slip-resistant mat could be used under the mixing bowl to help hold it in place while pupils are mixing. Alternatively, use a bowl with a suction base, a bowl holder or electric mixer. Make use of equipment with stands, such as an electric mixer with a bowl that is held, rather than a handheld beater.

9. A second handle on a saucepan will help with lifting and positioning the pan. Knob-shaped handles can be difficult to grasp. D-shaped handles are easier and can be lifted with the handle of a kitchen utensil. Some pupils may not be able to lift a pan of hot water easily and safely. If the hob is level with the adjacent work surface they should be encouraged to slide the saucepan off the heat. Use a wire basket insert to cook vegetables. This can be lifted out of the pan when the vegetables are cooked, making draining easier.

10. There is a wide range of adapted kitchen paraphernalia available, such as carrying trays, kettle tipping devices, kettles that change colour to indicate they have boiled, accessible ovens and grills, easy-to-operate can openers, jar stabilisers, cool-wall toasters, serrated knives rather than smooth-edged knives, different handled and shaped peelers, etc.

11. Choose equipment with larger handles or modify handles for pupils who experience difficulty in gripping, holding and manipulating equipment. One temporary adaptation is to place a slit tennis ball over the problematic handle. Polymorph plastic modelling material can be used to quickly make made-to-measure grips for individual hands.

12. Using an extra hand when using scissors, snips and cutters – clamp or fix one of the handles of the tool into a vice and clamp it to the work table. This leaves just one handle to operate; co-ordination and accuracy improve quickly and the downward pressure on the one free handle

increases cutting efficiency. Remember to remove the tool from the vice when not in use.

13. Low temperature electric glue guns are safer than other glue guns and can provide a quick-joint facility.

14. A sewing machine needle guard designed to protect the needle during transportation can be used to reduce the hazards of machine use with some pupils.

15. A safe-cut electric modelling saw that relies on vibrating, rather than a rotating or sliding blade, can be used for cutting curved profiles and templates.

16. Low voltage or rechargeable power tools cut down the number of trailing leads and the risk of electrical injuries.

17. A roller, wheel or rotary guillotine can be used to cut paper, card and plastic and may be safer for some pupils than scissors.

18. When cooking, some pupils may find it easier to use a chopping device rather than a knife, as the handle is pushed down and the blades cut the food inside. A slip-resistant mat placed under a chopping board will help to hold it in place, or some boards are supplied with slip-resistant feet or suction cups on their underside. Some boards also include spikes to stabilise food. If pupils need to slice a loaf, there are bread guides that will help them to cut even slices, particularly useful if pupils have a visual impairment. A coloured board is also useful if a pupil is partially sighted as it contrasts with the food they are preparing. Some chopping boards have fold-up sides which make a shunt to assist with transferring chopped food into a saucepan or peelings into the bin. An integral handle can help pupils to position and carry the board.

Figure 3.9 Vegetable cutter

Figure 3.10 Star cutter

19. Use plugs that are easy to remove from the socket and fit them with grips.
20. Use CAD/CAM and controlled machines to improve the quality of the outcomes if appropriate, for example for computerised embroidery machines for sewing logos, plotter cutters to produce iron-on graphics. 3D printers, milling machines and routers are useful for producing complex shaped pieces and a bread maker or food processor is useful for mixing and kneading bread.
21. There are many machines worth investigating to enable all pupils to produce high quality products. Some examples are:

   - *Ellison Electronic Cutter* – This cuts out precision shapes, including nets, lettering, finger puppets and a massive selection of novelty shapes. The press knives will cut a variety of materials, including card, felt, foam, vinyl, etc. It is also an excellent introduction to industrial style production at Key Stage 3 or 4.
   - *Roland Plotter Cutter* – This produces line drawings, or if fitted with a knife for cutting, it can produce self-adhesive vinyl, card nets, cardboard engineering, or even self-adhesive wood veneers. Graphics to be stuck onto products, silk screen masks made for printing, and iron-on vinyl can be used to customise garments.

- *Roland Stika* – This is a small scanning and cutting device where even simple hand-drawn images can be turned into self-adhesive vinyl within minutes.
- *The Badgemaker* – This is a tabletop machine which can be used to produce a range of products based around the button badge idea. It is excellent for production line work.

Ofsted recognises the excellent use of ICT and CAD/CAM in D&T so that pupils with special educational needs, and in particular those with specific physical disabilities, are able to overcome barriers to make things accurately and to turn their ideas into high quality products – good enough to buy in the shops!

> A pupil in Year 4 with cerebral palsy who was confined to a wheelchair was unable to speak without the aid of a computer. He had only some control over arm movements, but was well integrated into a group of pupils making torches. He was helped by CAM software, and specially designed tools, such as a stapler, that enabled him to make an equitable contribution to the work of his group.
>
> (Ofsted, *Meeting technological challenges? Design and Technology in Schools 2007–10*)

There are several ways in which ICT can have a valuable role in relation to D&T for pupils with SEND. One way is as a support for communicating (such as using symbols), another is an aid for modelling ideas (such as graphics and CAD software, or spreadsheets) and another is in making activities using computer-aided manufacturing (CAM) equipment such as sewing machines, plotter/cutters and so on. One of the benefits of using ICT is that the quality of the product can be greatly enhanced as the pupils are not limited by their motor skills or held back by their difficulties in presenting their ideas in design drawings, or have to rely on the teacher for oral instructions. Pupils with SEND should have high expectations; they are surrounded by high-quality images and products and become highly motivated when they realise that using ICT means that they can produce 'professional' looking designs and products. Often, using ICT means that they are released from the constraints that hold them back and are allowed access to a whole new designing and making world.

(See Appendix 3.10 for a self-audit to support you in establishing an inclusive D&T classroom.)

4 Teaching and learning

Structuring effective D&T projects

Typically, a D&T scheme of work is a combination of a range of different projects or units of work used to teach the national curriculum, or the equivalent broad and balanced curriculum that is planned by the school. The D&T projects vary from school to school, according to staff interests expertise on site. Each project should have a clear focus on teaching particular objectives or outcomes and an appropriate topic or context should be selected to help pupils to engage with the content. Some projects are more appropriate for pupils with special educational needs than others because they are easier to adapt and differentiate the tasks and the outcome. (Example planning sheets are provided in Appendix 4.1.)

The examples of design and make assignments (DMAs) in this chapter have been taught by a number of schools and they have suggestions about how to adapt them for pupils with SEND. The range of projects were chosen against a number of considerations:

- They are popular amongst pupils with SEND: we asked pupils which projects they had found motivating and interesting.
- They are manageable for the teachers and pupils because the outcome(s) can be controlled so that the end product is of good quality, but also so there are genuine opportunities for the pupil to make design decisions and work independently.
- They provide a range of materials, contexts, experiences and products.
- They offer progression and differentiation.
- They are flexible to the needs of different schools.
- They require equipment and expertise from teachers which are normally available (teachers in special schools often work in one room, using a range of materials).

A checklist to use when considering projects at Key Stage 3

Across the range of projects for this key stage there are three important areas to consider.

That pupils consider:

Tomorrow's technology.
Intervening to improve quality of life.
Working as a team member.
Aesthetics.
Technical issues.
Social issues.
Environmental issues.
Industrial practices.
Present D&T – its uses and effects.
Past D&T – its uses and effects.

That planning includes units which:

Aid transition from Y6 to Y7.
Aid transition from Y9 to Y10.
Are short.
Are lengthy.
Are focused (focused practical tasks).
Are open (design and make assignments).
Involve designing but not making.
Involve making but not designing.
Involve product analysis.
Start at various points in the design process.
Are set in a variety of contexts.
Take account of gender issues in relation to D&T tasks.

That pupils are involved in designing and making:

'desirable' products.
'culturally, environmentally and socially defensible' products.

Moving rooms and teachers – carousel style courses

In the most recent Ofsted subject report (2011), it is highlighted that rotational courses or carousel systems (where pupils spend a few weeks in a specialist room, on one project with a teacher, before moving on to the next room/teacher/project) are probably not designed in pupils' best interests, it states, 'this method had disadvantages in disrupting learning, particularly for students with special educational needs and/or disabilities, who had difficulty in connecting different aspects of the subject. It also made the tracking of students' achievement problematic' (Ofsted, *Meeting technological challenges? Design and technology in schools 2007–10).*

Most pupils benefit from routine, stability and a teacher who has the opportunity to get to know them over an extended period of time and can support effective differentiation as a result.

Selecting appropriate parts of the programme of study

Whilst keeping expectations high for all pupils, some parts of the programme of study are more appropriate than others. The teacher should have a clear rationale for what is prioritised and why. For example, Key Stage 3 national curriculum focuses on giving pupils opportunities to:

- suggest outline plans for designing and making;
- communicate design proposals;
- select and use tools, equipment and processes, including CAD/CAM;
- explore the properties of a range of contrasting materials;
- analyse products and judge the quality of others people's products.

Key Stage 3 expectations

While that is the teaching focus, not all pupils will achieve those things. However, given these opportunities, pupils should be able to:

- make choices about a product or aspects of its design, and;
- observe, explore and experience a range of materials and tools.

There are some parts of the programme of study that may be too demanding, for example, at Key Stage 3:

- generating design proposals;
- prioritising actions and reconciling decisions as a project develops;
- considering the chemical and physical qualities of materials;
- understanding systems and control or structures.

In this book there are some examples to show how you can adapt well known projects for KS3 for pupils with SEND. You might like to use this process to adapt your own favourite projects.

This process is best done with a colleague, departmental team or support staff, as discussion will be needed at each stage.

1. Identify the core learning objectives and outcomes (expectations) for the project – use the unit sections: expectations, learning objectives, learning outcomes. The key features given should remain.
2. Develop an overall view of the project and what is provided already and then list that aspects may demand further support for pupils. If you have already run this project with pupils recall the aspects they struggled with and which demanded extra teacher input. Look back in previous pupil files to see what they reveal.
3. Decide how this support may best be given, e.g. extra worksheets, a new piece of equipment, use of IT. You may want to present this as a chart (see possible problems/possible strategies).
4. Make a list of action points and a plan for what needs to be done.
5. Trial with your pupils and check readability and possible ambiguity of instructions, ease of use, etc.
6. Review – suggest improvements in the light of your experience.

Figure 4.1 How to adapt a scheme of work unit

This is not to say that pupils will not study these aspects, for example, *structures,* but that the Key Stage 3 programme of study expectation of pupils to 'calculate the effects of loads, forces of compression, tension, torsion and shear' was not appropriate. This point is not intended to ignore important aspects, but a simple recognition that it is better to focus on the most appropriate aspects of the programme in order to meet the needs of the pupil.

Units of work – examples of projects adapted for pupils with SEND

Using these principles, the following case studies demonstrate how to adapt the units of work (or projects) for pupils with SEND. It is essential that the units of work that pupils with SEND engage in are the same as for their peers.

Snacks project

Design and make assignment

Design a new and appetising filling for a pasty, which a target group of customers will want to buy. Develop your ideas by tasting other pasties and finding out what customers want. Experiment with different ingredients for the fillings, and test them to help you design something new and tasty for the customers.

Learning objectives

Pupils will be taught to:

- use a range of cutting, shaping and mixing processes;
- use a variety of techniques to prepare and process food;
- work hygienically in the classroom;
- use simple prototypes and modelling to evaluate design ideas.

The context

The teacher chose this project because pasties are a familiar product for the pupils and they have eaten them before. The task helps them to develop important manipulative skills. It focuses on designing just the filling and this makes it manageable. It can be extended to include the shape of the finished pasty. The project provides an opportunity for all pupils to achieve a good result. Ready made (fresh or frozen pastry) can be used.

As this project focuses on the filling preparation, it is possible to substitute a baked potato for the pasty.

Expectations

- Identify possible filling ingredients and mixes of ingredients.
- Evaluate the taste of a range of pasties which can be bought in shops.
- Improve preparation skills, such as cutting and rolling out.
- Select and use appropriate simple food-processing techniques safely.
- Consider safety and hygiene when handling food.

Some pupils may also:

- understand and apply their understanding of what happens when food cooks;
- develop ideas and problem solve creatively;
- work out the costs involved.

Hints and tips

Differentiating the focused practical tasks and managing the design and make assignment.

Some support sheets that the teacher used are available on the website: www.routledge.com/9781138714922.

- This can be a short project, over one or two lessons, or a longer project if pupils try out a number of ideas.
- Have a tasting and investigation session – 'What exists already?' – on the variety of pasties available. Identify the ingredients in the filling and how they might have been prepared (cut, grated, sliced, mashed). Talk about who would eat which pasty (introducing target market) and when it might be eaten (time of day/purpose). Record responses for the class. Some pupils can record their own responses (a pro-forma sheet might help), others can be assisted in recording – digital photos and key word prompts may help.
- Before pupils begin designing their own ideas, use a standard (prototype) recipe to make simple pasties as a class activity, led by the teacher step by step. These will help pupils to understand the limitations of what a pasty is like, what you can put in one and what you cannot. They will gain confidence and be more creative in their ideas. They will also be well equipped to plan their own making.
- Provide a recipe with pictures for pupils to follow and reinforce key words for equipment, ingredients and processes.
- To reinforce planning, cut the printed recipe into strips, thereby separating the steps, and ask pupils to put them in the right order.
- Minimise the number of stages required in a recipe by pre-preparing some parts of it. Use some ready-cut ingredients and ready-cut circles of pastry. It is important not to do too much for the pupils, but this will depend on their ability.
- Templates can be used to make pastry circles – such as a small plate. Uneven rolling out can be prevented by using blocks which have been made for this purpose to the required thickness and balancing the rolling pin on them.
- Understanding how much filling will go into each pasty can be a difficult concept and will require the teacher to guide the pupils.
- Spend some time exploring and explaining changes in food when it is cooked, using the correct terms where possible (melts, softens, hardens, browns, shrinks, etc.). Challenge pupils to apply their understanding to their own designing by asking them questions: 'Why are you cutting the onion so small?', 'What happened to the cheese?', 'Should you grate the cheese or cut into cubes?', 'Why do you use minced beef?', 'Which will look better?'.

- Inspire pupils with a collection of real or photographed ingredients which they could use for the filling. Challenge them to name them by providing the labels to put on.
- Encourage pupils to develop their designs and make modifications where appropriate. They can record the recipe afterwards.
- Provide pro-forma sheets, or a spreadsheet, to work out the cost of the pupil's recipe. With some pupils you can compare the cost of the recipe and the cost in the shops and talk about manufacturing costs and profit.

Fabric case/wallet project

Design and make assignment

Design and make a wallet or small fabric case for a particular purpose, such as to hold coins or credit cards, to hold keys or to hold pencils. Develop a standard prototype and then show how it can be adapted simply to make a number of new products for different users, e.g. by using decoration, different fabrics or fastenings, and how it can be batch produced.

Learning objectives

Pupils will be taught:

- batch production, and how basic design can be personalised for a particular person or client;
- how to use a template/pattern.

The context

The teacher chose this project because each pupil achieves a high-quality product that they can make for their own use. In addition to the pleasure of making something to keep and use, the project helps pupils to understand the properties of textiles, and the ways in which fabrics can be constructed to fulfil a purpose (such as protection). There is an opportunity to teach valuable construction skills – using patterns, pinning, cutting out fabric, sewing by hand and machine skills. It is further enhanced when the pupils are able to use CAD/CAM embroidery machines to personalise the fabric case/wallet as this will give it a professional finish.

The task can be extended to include work on knitted and woven fabrics, and fabric tests such as absorbency and stretch, as the results are visible and concrete.

In order for the making aspects to be manageable, the pupils are limited to one style of fabric case/wallet. A pattern template is provided for this and all the pupils will be making the same style of fabric case/wallet after the class agrees on the design. With the style agreed, pupils will focus on other design decisions such as colour, decoration and fastening.

Expectations

Pupils are expected to:

- Evaluate existing fabric cases and holders in a product evaluation.
- Choose the colour, fabrics and fastenings.
- Design their own personalisation to decorate the fabric case.
- Model the design in paper or inexpensive fabric before making so that they can then plan the order of making and test out their idea.
- Use a sewing machine assisted by a helper where needed.
- Use a simple pattern.

Hints and tips

Differentiating the focused practical tasks and managing the design and make assignment.

- Provide a range of fabric cases for the pupils to evaluate. For example, ask pupils to bring fabric purses or pencil cases. If you are limiting the group to making one product, take care that pupils do not think that they can make other types of cases. Talk about: 'What materials are used?', 'How is it fastened?', 'How is it finished?', 'How does it protect the contents?', 'How does it organise the contents?', 'How many parts are there?', 'How is it assembled/in what order?', 'How big is it; is it easy to use?', 'How would you recognise that it is yours?' – and record the responses for the class. A digital camera and pro-forma sheet can be used. Record key words and information that will help pupils generate design ideas.
- Spend some lesson time making simple items in order for pupils to practise using a sewing machine to produce straight seams and hems. Some pupils will be able to learn how to set up a sewing machine; providing pictorial instructions will help them. Reward pupils who can do this with a class mentor badge or certificate.
- The teacher should demonstrate and advise pupils on how to make seams and hems. Depending on their previous experience, pupils will need to practice these as you need to be skillful with the sewing machine to do them. Time spent practising straight lines will likely be valuable to pupils and some will need assistance to sew. However, the adult should always act under the instructions of the pupil.
- Have the pattern templates ready for the pupils to cut out.
- It is important that the pupils work out the sequence for making the case. Putting the decoration and pockets on the main part of the container before

sewing it together are key concepts. A jumbled-up flow diagram of the stages, with pictures and key words, would be very helpful.

- Transferring the seamline (sewing line) from the pattern to the fabric with chalk, transfer wheel or tacking, would be vital for pupils with coordination difficulties.
- It may be helpful, depending on the ability of the pupil, to limit the choice of fabric to ones that fray as little as possible (fleece and PVC are very good options).
- Provide a limited selection of fasteners for pupils to choose from.
- Pupils will need help with designing the logos. Simple scan and sew machines are helpful, and stencils and inkjet transfers can be used if no computerised sewing machine is available.
- CAD/CAM embroidered designs can be embroidered onto a piece of fabric which is sewn onto the fabric of the case/wallet. This allows pupils to improve their design and mistakes on the embroidery will not affect the whole product.

Colouring textiles project

Design and make assignment

Develop your own colourful design on fabric and use it to make a bag, T-shirt or wall hanging. Investigate some of the ways of colouring fabric – tie and dye, batik and block printing.

Learning objectives

Pupils will be taught to:

- identify suitable materials and techniques, taking into account appearance and function;
- name and describe the methods and processes used to colour fabrics;
- use decorative techniques for a purpose.

The context

The teacher chose this project because pupils can respond flexibly: it is adaptable to any school's circumstances and pupils' ability levels. It develops important decorative and creative skills.

Pupils may achieve better results with this design and make assignment if the outcome is limited to a particular product (such as T-shirt, shorts or a bag). The emphasis is then on learning about different colouring techniques, and applying this knowledge to enhancing their own designed product.

Expectations

Pupils are expected to:

- Explore colours and their meanings.
- Explore a range of colouring techniques.
- Choose one technique, or a combination of techniques, to use on their product.
- Choose colours and colour combinations for their product.

Hints and tips

Differentiating the focused practical tasks and managing the design and make assignment.

- This is a great opportunity to explore colour in fabrics! Spend some time exploring colour on different textiles products with the pupils. Talk about the different colours, how the colour has been transferred to the fabric and the patterns present. It is an opportunity to link with art, look at colour wheels and explore colour mixing. Colour boards for each colour could be developed to produce a classroom display. If there is time, pupils could also work with natural dyes from berries and leaves they have collected.
- Developing awareness of colours can be achieved by:

  - weaving with coloured paper to find good colourways;
  - wrapping different colours of yarn around card to see what effects can be achieved with a limited range;
  - using the computer to produce quick effective moods/atmospheres on the same image;
  - using wrapping paper/wallpaper fabrics to see why colours are good together.

- Use examples to get pupils thinking about the meanings of different colours.
- Spend some time exploring the techniques first. Pupils may start this project by exploring and experimenting with the different colouring techniques. The making of samples is a motivating end product in itself. It is helpful to record the steps using a digital camera to reinforce learning later. It may be found that if pupils know the 'product' too soon, they don't explore as readily. Once they have practised batik, tie & dye, and block printing, they could design using the techniques for the range of outcomes – be it T-shirt, sarong, bag, etc. – and then select which they prefer and which they're most successful at.
- The focus is on simple techniques which the pupils must organise well.
- Have some examples of decorated products on display, clearly labelled with the technique used, for discussion and evaluation and to get ideas.

- Choose fabrics carefully, as synthetic fibres resist natural dyes.
- Tjantings are quite difficult to use. Make wax paintbrushes by attaching foam to pea sticks with wire.
- When tie dyeing, use elastic bands instead of string to make tying easier.
- Steep the ends of garments in a tub of dye and allow pupils to watch the colour creep up the fabric.
- Use a ready-made T-shirt/bag if time is limited or the making skills are covered elsewhere.
- Plan the decoration of the final product by talking about it as a group.

'Keeping it under control' project

Design and make assignment

A stage play, a pop concert, an animated story or a puppet theatre all involve performances. Presentations to attract people to an event or advertise a product can use animation and other performance techniques. Many museums and theme parks have interactive displays. The best way of producing these effects is to use different types of control systems.

Your challenge is to combine mechanical, electronic, and electrical systems to produce a working model that could be used in a performance or presentation.

Learning objectives

Pupils will be taught:

- how to work as a team to produce a performance;
- how to produce a working model;
- how systems require control and how this can be achieved;
- basic principles of control systems;
- how to make structures strong.

The context

The teachers chose this project because they were confident in the area of control and wanted a flexible and easily adaptable project for different needs that would motivate pupils. It is possible to run it as a group project. It provides important coverage of *systems and control* and *structures* aspects of the D&T curriculum.

Systems and control is a complex area with some abstract concepts that make it more difficult for some pupils. Some projects such as 'keeping it under control' can be open-ended so the teacher must set limits on the outcomes expected from the pupils, while at the same time allowing them some freedom to explore their own ideas.

To make the project meaningful it can be put into a relevant context, for example:

- The puppets could be used to provide a performance for other pupils or young children.
- The toys could be for a particular young child or for a playgroup.
- The display could be for use around the school, or in a community building like a local library or a museum. It could be used to provide information for a particular group of people or in a particular location.

Expectations

Pupils are expected to:

- Recognise simple electrical and mechanical control in existing products, such as a lever in a pop-up book.
- Construct simple electrical control circuits that include switches and outputs in parallel and series, protection of LEDs and reversing control of motors.
- Explore four different kinds of motion and how mechanisms can be used to change one kind of motion into another.
- Use cams and linkages.
- Model a strong structure.

Hints and tips

Differentiating the focused practical tasks and managing the design and make assignment.

An example of how lessons can be differentiated according to pupil level is given in Appendix 4.2.

- Use a series of pictures or examples of models/products that pupils can make. This should lead to each pupil coming up with a realistic and manageable task for themselves.
- Pupils may believe that anything to do with control systems is too difficult for them to understand. They need to be shown simple examples of control systems and products. Use demonstrations and display of a range of 'control products'; mechanical and electrical toys, puppets, labour saving gadgets, etc. for them to consider. Visit a museum and/or interactive science centre.

- Pupils may need help with planning, as they may find it hard to see the need to work in a sequence of stages which need to be planned in advance. Break the project down into smaller steps. Use a planner with key words and pictures of the main stages to help pupils to order the sequence of actions. In systems and control, the key planning should include working through the following sequence and discussing the possibilities with pupils:

  - What do I want to happen? (Output)
  - How can I make this happen? (Device to use)
  - What do I want to use to cause it to happen? (Input)

- Use electronic systems kits, mechanical kits, and other pre-manufactured structural components to help pupils achieve the desired outcome quickly.
- When making, some pupils will find it hard to work to the level of accuracy required, e.g. making mechanical linkages, or assembling electronic circuits. Make use of as many standard components as possible. Use a well-labelled container for electronic components (use pictures) and paper soldering equipment including clamps.
- Fault finding in electronics requires patience and a systematic approach. Show pupils how to modify things that do not work well first time, and encourage them if they become disheartened. Explain that this is part of the *process* rather than being a failing on their part!

Sheet material project

Design and make assignment

Many products are made from a single sheet of material, e.g. flat sheets can be folded to form a 3D shape to make furniture, containers or packaging. Objects made from folded sheets can be very light and rigid, and folding flat sheets to form a 3D object can be more economical in production than joining separate sections. Complicated shapes can be modelled using card or computer software.

Design and make a product that is primarily made from a single sheet of material. Your product should be aimed at the young teenage market and easily batch produced.

Learning objectives

Pupils should learn:

- that products are designed to meet particular consumer needs;
- how to use a range of cutting and shaping processes, selecting and using

appropriate hand tools to cut, join and shape specific materials safely and accurately;

- that manufacturing aids, e.g. jigs and templates, ensure accuracy and help with volume production;
- to choose suitable materials and an appropriate construction method.

The context

The teacher chose this project because it is easy to relate to pupils' needs and lives. It covers important practical skills and utilises a range of sheet materials. It gives pupils a good foundation in working with a range of materials and processes and a solid understanding of structures, but it can be easily adapted to be challenging and demanding at an appropriate level. It could be a group project.

This project involves pupils designing and making products in sheet plastic (probably acrylic), sheet wood and sheet metal. They explore the properties of materials. One approach is to show the pupils some examples of products they can make. Everyday products are best for introducing this project. Depending on ability, the teacher can limit these to a simple product, such as a letter rack, desk tidy or clock. This will make it easier to manage but still give the pupils some choice and control over the entire design.

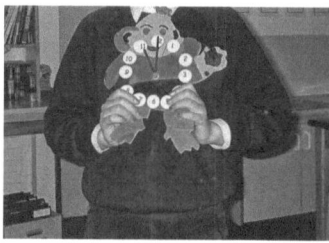 

Figure 4.2 Clocks

Expectations

Pupils are expected to:

- Generate their ideas by considering the properties and features of existing products.
- Model their ideas using card.
- Explore how sheet material behaves.
- Choose the sheet material and technique for cutting, folding, joining and finishing.
- Assemble their own product.

Hints and tips

Differentiating the focused practical tasks and managing the design and make assignment.

- Ask the pupils to work out how the products are made, why they are made that way and what materials are used. Guide discussion where necessary.
- Use a planner sheet with key words and stages to get the pupils started on their design work and to help them plan their designing and making.
- Encourage straight line bending and circular holes to ease production.
- Make some jigs and formers in advance.
- When working in acrylic, always drawer file, scrape, sand wet and dry with 800 grit sandpaper and buff to a shine.
- Layers of thin 1mm plywood can be glued together while being held in shape by a former to produce very interesting and strong shapes.
- Health and safety – the edges of sheet metal and other materials may be very sharp: always make sure pupils file a safe edge before they start working the metal and after any cuts are made.

'Puzzling boxes' project

Design and make assignment

Many small gifts, games and puzzles are produced to appeal to a particular group or type of person, e.g. 'Kinder eggs' contain a surprise gift for young children inside a chocolate egg.

The BB Box Company has asked you to make a box of a maximum size 100 × 100 × 100mm. They would also like you to design an interesting gift and finish it in a suitable way.

Learning objectives

Pupils learn how to:

- use simple prototypes, mock ups and models to evaluate design ideas;
- use a range of cutting, forming and shaping processes, e.g. sawing, line bending;
- use specified hand tools to cut and form material safely;
- use CAD/CAM to make a puzzle aimed at a particular user.

The context

The teacher adapted this project easily so that it was very appropriate for pupils with SEND. A simple box is made to a given size. The whole group of pupils can make the same box, and this aspect of the project can be teacher led, step by step. While the box is limited, designing puzzle and decoration provides opportunity for differentiation. Pupils have the opportunity to design and make a high-quality finished product that they will be proud of, by using CAD/CAM machines to design and make their puzzle. However, it is possible to do this project without a CAD/CAM machine if schools do not have these facilities, as hand tools may be used in the absence of a CAD/CAM machine.

Expectations

The pupil's design input is focused on being able to:

- cut the lid to suit the chosen purpose of the box;
- finish the outside of the box appropriately for the person/purpose for which they are designing;
- develop a puzzle to fit inside the box that is appropriate for the person/ purpose.

Hints and tips

Differentiating the focused practical tasks and managing the design and make assignment.

- The size and the shape of the box that the pupils will make will be a major constraint on the type of puzzle that the pupils design and make to go into it. A successful approach is to make the box first with no reference to the puzzle, simply telling the pupils that the box will be needed for the next part of the project. Being able to see the box has been shown to help many pupils in the design of the puzzle.
- The puzzle box is made as a rectangular or square tube with ends on, which is then cut into two 'halves'. The dimensions of the box can be carefully chosen so that it can be cut in two with a band saw or Hegner saw (the power saw to be supervised by the teacher at all times). Birch plywood is the preferred material because of the range of finishes that can be obtained. MDF is also suitable and has some advantages in cutting and sanding. The use of 6mm ply for the two sides and 4mm for the rest has proved valuable. If glue is applied to the 6mm ends, greater surface adhesion is obtained. It has been found satisfactory to use 1.5mm ply for the end pieces since it is very easy to trim and sand to a good finish.
- By using different cuts and ink stains, a highly constrained box becomes a unique product. Have a few examples ready to show the pupils.
- Make a few first yourself and explore different things such as the type of finish, the number of inserts, different puzzles, etc. This will give you an

understanding of the difficulties that may occur and provide a range of examples that may be used to motivate the pupils.

- When the lid has been cut off, hardboard can be used to make the inserts. One on each side is best, but two on opposite sides will do.
- Introduce the puzzle part of the project by asking the pupils to evaluate a variety of puzzles aimed at children and adults. Ask them who they have been designed for; What is the purpose of the puzzle?
- Use lightweight card to model the puzzles and try them out in the box.
- If you do not have a CAD/CAM machine you should still design the puzzles on a computer, then print the design, cut it out to make a template, and stick it to the material being used to make the puzzle.
- Don't be put off by the CAD/CAM aspect of this project. If you have CAD/CAM machines it is an ideal way to get started, if not then the alternative method mentioned above really works well and still gives pupils a good, worthwhile experience.

Classy casting – art deco jewellery project

Design and make assignment

Art deco is the name given to a style which was in fashion during the 1920s–30s. Designs from this period are usually based on combinations of very simple geometric shapes.

Using pictures of art deco designs, create an attractive piece of jewellery which is up to date enough to be worn on a special occasion.

Learning objectives

Pupils will learn:

- how designers rework old and existing ideas to come up with something new;
- about an important period of twentieth-century design history;
- how metal casting can be used, specifically how cuttlefish moulds are used to cast pewter.

The context

The teacher chose this project because it encourages pupils to use source material to develop simple ideas. While drawing skills are included, they are not necessary as pupils can design into the mould. Moulds produce high-quality outcomes which are quite easy to achieve. It introduces, in an effective way, principles of casting in metal and industrial manufacturing. Pupils are proud of their jewellery at the end of this project!

The context of art deco requires pupils to look at existing designs to help them design something new. The ability and motivation levels of pupils will determine the depth of study and the time spent on looking at design style and generating their own designs.

Expectations

Pupils are expected to:

- Use art deco pictures to create an idea for a shaped piece of jewellery.
- Carve the shape into the cuttlefish bone mould.
- Make the shape by casting with pewter and attach the findings.
- Clean, buff and polish the jewellery.

Hints and tips

Differentiating the focused practical tasks and managing the design and make assignment.

- Have a range of source material covering the art deco period to show to pupils.
- Show the pupils how to go from a picture in source material to a sketch of an idea, for example by simplifying a shape (outlining), or by masking (part of a shape, or magnifying). Scanned images are useful for simplifying.
- Bold shapes will lessen problems with filing and polishing. Intricate shapes are likely to bend and snap.
- It is possible to go straight to the modelling stage using card or playdough to help pupils to develop their ideas if they are not good at sketching. The modelling helps pupils understand how they will need to sculpt their mould. The parts of the model that stand out the furthest will have to be cut the deepest.
- A plan of the process step by step will be helpful.
- The teacher will need to prepare the cuttlefish bone. Any good pet shop will let you have a box of cuttlefish bones at trade price. Unfortunately, it can make the storeroom a bit smelly. When cutting the top off the cuttlefish bone, use a hacksaw and cut from the hard side. This prevents the cuttlefish bone cracking and splitting. (Every cuttlefish bone has a hard side and a soft side.)
- Instead of using cuttlefish bone, layers of hardboard can be used.
- Flatten the soft side of the cuttlefish bone for sculpting the mould. This needs to be done using sandpaper. Don't rub back and forwards because it will result in a curved surface. Put the cuttlefish bone on the sandpaper and pull it towards you and repeat until one side is flat.
- Show the pupils how to sculpt the mould with a used ballpoint pen and shake the dust off as they go. Protective glasses should be worn. Using a

ballpoint pen to sculpt will give good detail as it is easy to control. Don't touch the flat surface or the sculpted mould shape with your hands because it is so soft that the shape can easily be damaged.

- Pewter has a comparatively low melting point so you do not need a blazing hearth. If you have a dedicated casting/heat treatment area, it is best to use it. Small gas stoves designed for camping, and even domestic hobs, have been used but health and safety issues must be observed. A large biscuit tin with dry sand in it can be used to stand the cuttlefish bones in for pouring. This provides a safe environment for any spillage and the spill can be reclaimed when cold. All modern pewter is lead free, old pewter is not.

Corporate identity project

Design and make assignment

Souvenirs and collectables, such as T-shirts, 3D signs and models, are used to promote events, pop stars, cartoon characters and even schools.

Design and make a coordinated range of promotional products for a special occasion or a corporate client. You should work in a team and produce at least three different products using a range of materials.

Learning objectives

Pupils will learn:

- to use ICT effectively when working on a collaborative project;
- to use ICT to research and analyse information and expertise from outside the school and to collect appropriate information for their project, such as examples of logos;
- to develop strong teamwork skills;
- how to use CAD/CAM.

The context

The teacher chose this project because a range of outcomes are possible and individual pupils' responses are easy to accommodate. Corporate and promotional products are easy to source and generate many pupil ideas. The teacher can focus the project on a real class or school event to give the pupils context to which they can relate. It is also possible to make the research and design of the logo or corporate image a whole-class designing activity. The pupils then individually use the design to make their own products. The same logo/image can be printed, embroidered or milled onto products according to the ability of the pupil. The teacher restricted this activity to produce a banner or T-shirt, but pupils made many of their own design decisions.

Expectations

In bringing together what they have learned to make a promotional T-shirt or banner, the pupils are expected to:

- Choose an appropriate image and text to be used.
- Choose the colours they thought most appropriate to the task.
- Choose the method of transferring the image and text onto the T-shirt or banner.
- Use an on-screen grid to select words and symbols.
- Use a computer sewing machine (with help).
- Practise screen printing and use fabric crayons/pens (with help).

Hints and tips

Differentiating the focused practical tasks and managing the design and make assignment.

- Have a range of promotional T-shirts and other items with designs on them to investigate and prompt discussion.
- Show the pupils how to use, and let them practise using, a digital camera.
- Show the pupils how to use, and let them produce, images/drawings using computer software.
- Show the pupils how to use, and let them produce, screen-printing templates using a printer.
- Show the pupils how to use, and let them practise using, a word processor to produce writing and symbols.

'Feeding special groups' project

Design and make assignment

People demand choice, whether they are eating a meal on an aircraft, in a restaurant or in a motorway service station, and meals and dishes have to be made to suit a range of different dietary needs.

A company wants to buy a new meal to offer to customers with special dietary needs. Design a prototype of a suitable dish and present it to the company.

Learning objectives

Pupils will be taught:

- what is meant by a special dietary need, such as vegetarianism;
- how foods contribute to a healthy diet;
- how to select appropriate ingredients according to their nutritional characteristics;
- how to combine ingredients for specific purposes;
- how to work safely and hygienically.

The context

The teacher chose this project because nutrition is an important area to cover in food technology. It is often of personal interest, as pupils with SEND often have special dietary requirements. This project also promotes an understanding of good hygiene and food safety practices.

A focused project was used, which looked at the dietary needs of vegetarianism. The teacher can choose an appropriate dietary need for the class, rather than allowing pupils the choice to research any of them. This eases management issues which might be encountered if pupils all undertook different dietary needs and ensures that successful products can be designed and made by all pupils. The meal could also be developed by the group with individual pupils contributing different parts of the meal.

Expectations

Pupils are expected to:

- Explore, taste and comment on different dishes and recipes that are available for special diets.
- Choose ingredients appropriate to the special diet and *The Eatwell Guide*.
- Assemble appropriate, specific ingredients to make a recipe for a special diet.
- Be able to explain why they have chosen the ingredients.

Hints and tips

Differentiating the focused practical tasks and managing the design and make assignment.

- Provide a supportive structure and show the pupils a manageable way through the tasks. Segment the activity into smaller tasks with specific targets.
- Keep this project short if required.

- Evaluate existing vegetarian products to give pupils ideas to help them with their designing. A selection of different types of vegetable burgers could be used for the sensory evaluation test. Alternatively, a range of vegetarian meals, e.g. 'meals for one', could be used to help introduce the ideas of portion control and a 'balanced meal/diet'.

- Investigating products designed for vegetarians makes a good homework activity for some pupils. Pupils could look for the vegetarian symbol. Show a variety from different packaging from a variety of retailers and provide some information sourced from the Vegetarian Society. An extension of this could be a group activity to produce an information sheet, wall chart or class display for the school on vegetarianism.

- Reduce the amount of reading, researching and written recording required. Simplify and adapt research information, provide key words and pictures. Discuss different dietary needs with flashcards of different people/situations, for example show a picture of breakfast: 'What would a vegetarian have for breakfast?' If possible, have real alternatives/ingredients for the pupils to point to or talk about.

- If an outside visit cannot be arranged, ask the school meal service or the community dietician to visit the class. This may help the pupils to develop a questionnaire and/or a new item for the school lunch menu. Visiting speakers should be briefed carefully.

- A visit to the school kitchen may enable pupils to see 'hands on' the storage areas, health and safety procedures and equipment, such as temperature probes (for pupils to understand food safety).

- Include lots of focused making tasks led by the teacher to a set recipe; this will build up the pupils' skills.

- Encourage the pupils to choose recipes where they will produce good quality outcomes, so that they build their self-esteem and confidence.

Soups and salads project

Design and make assignment

Health experts recommend that we eat *at least* five portions of fruit and vegetables each day to keep our bodies healthy and working properly.

Design and make a new salad or soup which looks good and will appeal to teenagers.

Learning objectives

Pupils will be taught:

- how to classify ingredients, for example by food groups;
- how to prepare, cut, mix and heat ingredients safely;

- where fruit and vegetables come from and what they are called (and what their parts are called: roots, stalk, etc.);
- how to find out what teenagers like.

The context

The teacher chose this project because it is easy to adapt as a number of simple, to increasingly complex, products are viable options. It provides an opportunity to teach basic food concepts and skills. It is a short project, that can be extended if required.

Expectations

Pupils bring together what they have learned to design and make a salad or a soup/drink for themselves. They are expected to:

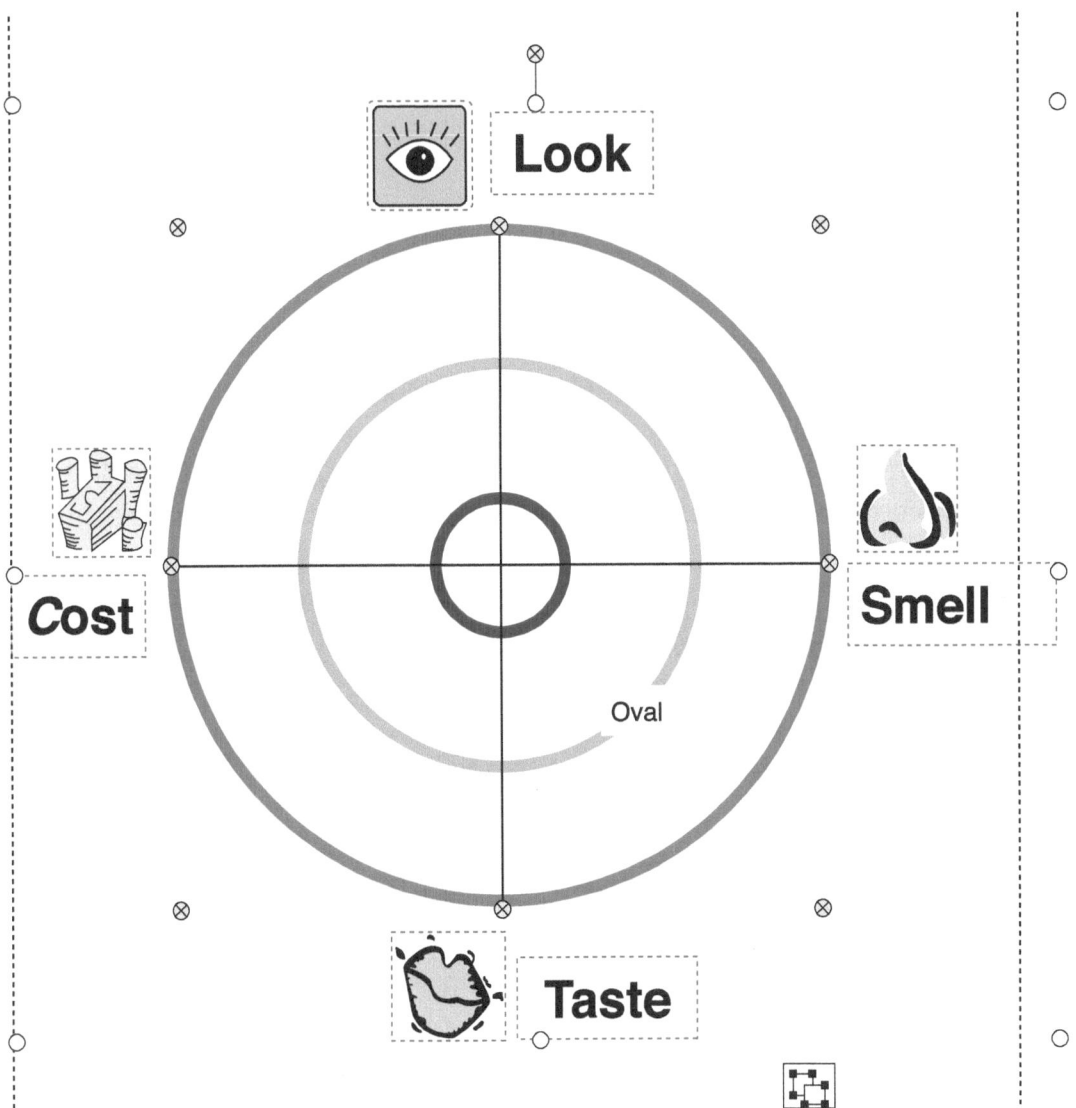

Figure 4.3 Evaluation grid

- Choose from fruit and vegetables provided by the teacher.
- Choose how to prepare and cut them.

Hints and tips

Differentiating the focused practical tasks and managing the design and make assignment.

Provide opportunities for pupils to *investigate familiar products* and explore materials. Pupils should:

- Observe, touch, smell and taste different fruit and fruit juices and respond to them.
- Record as a class how many pieces of fruit they eat each day.
- Collect pictures of fruit and vegetables and label them; use them for choosing ingredients when designing; put them into groups (colours, juicy/not juicy, etc.).
- Observe fruit/vegetables growing.
- Observe, touch, smell, taste ready-made products made from fruit/vegetables (for example, canned fruit salad, chilled salad) and respond to these products.
- Observe and explore changes in fruit (for example, the teacher shows them a raw apple and a cooked apple; the teacher leaves a cut apple to go brown).
- Teach essential practical skills through *focused practical tasks* led by the teacher before the pupils think about designing:

  - With help, pupils wash, clean, peel, cut, slice, grate, squeeze and mix different fruit/vegetables and fruit juices, such as making coleslaw or fruit smoothies to a recipe set by the teacher.
  - Hard foods should be chopped or sliced while being held in place with a fork or other holding device.

- Use serrated knives rather than smooth-edged knives.
- There is a wide variety of different handled and shaped peelers, etc.; be mindful of the pupils' needs when selecting them.
- Use some (but not all) pre-prepared ingredients for some pupils.

When supporting designing, provide simple choices from a given range when working relative to the ability of the pupil. Allow pupils to design 'as they make' and make changes as they go along. Allow opportunities for pupils to try again and modify recipes, as this is quite a quick product to make.

Moving stories project

Design and make assignment

Design and make a storybook or display that has moving parts for a particular purpose. (Individual teachers add contextual information according to the project they are doing, for example, making a story prop which includes a winding mechanism; or design characters for our story so that they look good, can be seen across the room and are easy to attach to the 'winders'.)

Learning objectives

Pupils will be taught how:

- products with lever and linkage systems function;
- to use particular mechanisms for a specific purpose;
- to use technical vocabulary to describe the properties of materials and mechanisms, e.g. 'lever', 'pivot' and 'linkage'.

The context

The teacher chooses a simple mechanisms project for the class by drawing upon suggested resources from an earlier key stage unit, where appropriate for the pupils. The focus of the project is to introduce pupils to the concept of winding mechanisms, building on their knowledge of wheels and axles. They explore how to make winding mechanisms using construction kits then, after discussion, make their own, using appropriate materials.

Expectations

Pupils bring together what they have learned to design and make a winding mechanism for a story or display.

- They work as a class, choosing the story, and each child makes a specified part of the design, with appropriate help.
- The teacher provides a template, a choice of designs, or allows the pupil to design their own part, according to their individual needs.

Hints and tips

Differentiating the focused practical tasks and managing the design and make assignment.

- Provide opportunities for pupils to *investigate familiar products* and explore materials; they may:

- Observe and explore simple winding mechanisms in a range of toys.
- Observe how the mechanism works and respond to the movement of the mechanism.
- Become aware of the different parts, such as wheel, axles, winder, pulley and gears.

- Teach essential practical skills through *focused practical tasks* led by the teacher before the pupils think about designing. The pupils may:

  - Build a winding mechanism using construction kit parts (with help).
  - Practise controlling the mechanism and become aware of cause and effect (if they turn the winder, the pulley winds the cord around it and the hook rises).
  - Explore ways to pick up items such as using a bucket, magnets or a hook.
  - Measure and cut small-section timber or dowel (with help).
  - Assemble parts, such as joining axles to base, and wheel to axles (with help).

Case studies reflecting examples of good practice

These teachers and schools in the following section all:

- Select projects where it is known that pupils will achieve a good end product in a context that motivates them.
- Have clearly focused objectives for the whole class and each pupil.
- Differentiate appropriately to meet the needs of individual pupils. Have different expectations for individuals within a group from the outset, with planned extension work to challenge some pupils and adult and peer support for others where needed.
- Provide a full range of activities, such as focused practical tasks, product evaluation, and design and make assignments.
- Aren't afraid to direct and structure some aspects of the project, but plan opportunities for pupils to be responsible for certain aspects of designing and decision making. This allows as much independence as possible.
- Break down larger projects into smaller achievable stages. There are many mini-making activities for pupils to motivate them and build up their confidence.
- Do much designing actively with real materials: there is less reliance on drawing and presenting ideas on paper.
- Use cross-curricular links and learning from other subjects or topics effectively to save time and enable pupils to apply learning in other contexts.
- Use ICT to support learning and to provide a means for recording and presenting work more quickly so that time is gained for other activities. (A template for a project proforma is supplied in Appendix 4.3.)

Case study: Working Together Week – Developing partnerships

This case study is about an all-age MLD (moderate learning difficulty) school which caters for 150 pupils. They say that Working Together Week was probably the most exciting week in their history. Thirteen schools took part in a multicultural activity week – five secondary, one secondary PRU, five primary, one independent and one special school. The regular timetable was disbanded and many exciting projects were organised for pupils. Experts from a range of backgrounds came into school to lead activities. Staff from each curriculum area planned activities with a multicultural focus for delivery throughout the week.

The aims were to:

- build bridges and open opportunities for racial harmony;
- develop tolerance and understanding;
- increase knowledge and experience of a variety of cultures;
- break down barriers within the community that existed around a special school, through teamwork, parental involvement and with mainstream pupils;
- offer stimulating and exciting learning opportunities, by introducing professionals from a variety of cultural backgrounds, with particular areas of expertise;
- offer an opportunity for teamwork in inclusive groups, culminating in a celebration of our diversity.

This was a whole-school activity involving all age groups (5–16 years). A variety of artists were employed for the week:

- an Afro-Caribbean story-teller;
- an Afro-Caribbean musician, actor, and story-teller;
- an Asian hand painter (Mehndi);
- an Asian artist (Rangoli coloured rice pictures);
- a textile artist;
- an Indian dancer (Bhangra).

In addition, staff from each curriculum area planned activities with a multicultural focus for delivery throughout the week.

How was it organised?

Each day, 48 pupils arrived – 36 primary and 12 secondary. The schools taking part made sure that the event was high profile – in some cases, pupils had to write a letter of application for a place.

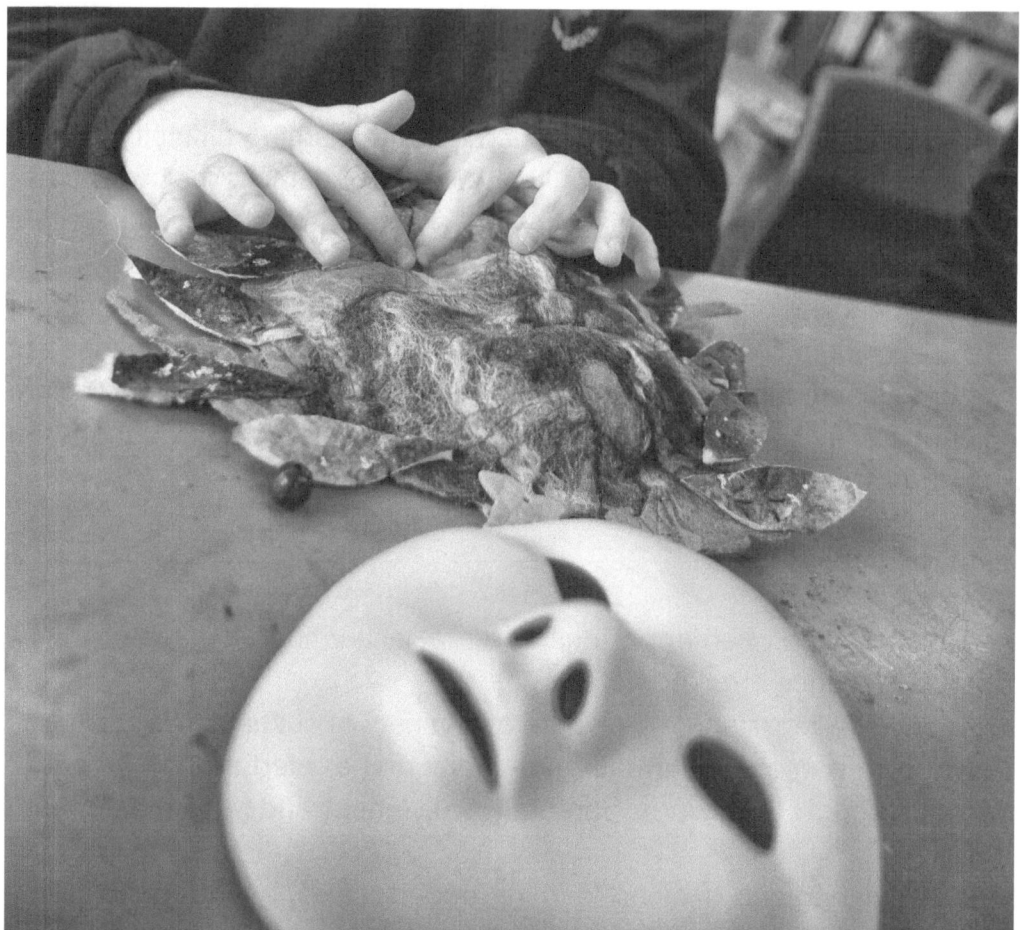

Figure 4.4 Masks

Usual class groups were re-organised and 12 new groups were created. Mainstream pupils were added to every group and pupils were discouraged from wearing uniform so that it was hard to tell where they had come from.

The week ended with a morning at the local civic hall, celebrating the diverse activities that had taken place. Through exhibiting and sharing work, children were able to demonstrate what they had learnt about other cultures and values during the week.

A colourful array of design-related work was produced during the course of the week. Individual achievements included mobiles, masks, Indian headdresses, ceramic jewellery, and computer-designed T-shirts, place mats and coasters. The pupils made their view of the top of the world (the Arctic Circle) into a 3D model. They used mathematical skills to produce Paisley patterns using rotational symmetry, they listened to stories from different countries around the world and they designed book covers for a competition. The project was deliberately planned to ensure that the skills, experience and expertise of people from local ethnic-minority communities were utilised within the curriculum and ensure that they served as positive role models to the pupils.

Working as a team

The collaborative projects provided a number of permanent mementos of the week, including:

- a beautiful wall hanging made from textiles, produced using a range of processes;
- a frieze depicting a story, made in a day by primary children;
- a batik banner made up of 48 sections by different pupils;
- clay tiles which, when fired, will be mounted as a mosaic.

Each school that was involved in the project received a copy of a newspaper about the week that was compiled by primary reporters.

The pupils had the opportunity to work with people from other cultures.

They:

- developed an understanding of different lifestyles;
- learned about differing dress styles, eating habits and beliefs;
- recognised, respected and valued the diversity of humanity;
- recognised that every group embraced a range of talents;
- developed friendships between mainstream and special pupils;
- recognised and celebrated achievement at all levels.

Throughout the week, all of the pupils involved in the project learned new things, had fun, made new friends and grew in confidence with each other. Evaluations received from pupils have been extremely positive and more than justify the significant amount of effort and planning that went into the event. Parents have commented on how animated and excited their children were when they came home (not to mention exhausted!). Some said that their child talked more that week than ever before.

When asked, 'What was the most important thing you learnt?', pupils commented:

- 'No matter what school people go to, or what may be wrong with someone, everyone has a life and feelings to share.'
- 'Working with others and respecting their thoughts and views.'
- 'How to make friends with people from different schools.'

Following up after the week, here are some conclusions:

- Staff have had the opportunity to experiment with multicultural activities relating to their subject areas. They can now build them into their schemes of work.

- The week should be repeated, but not as an annual event as first envisioned; it will possibly be held every three years.
- Staff have benefited from the experience of the visiting artists.

Case study: Linking with other schools

A special school approached a local mainstream school looking for opportunities for their 17- and 18-year-old pupils to work alongside pupils of a similar age in a different environment.

The aims of the project were for pupils of the special school to work alongside mainstream pupils for one morning a week. The groups were introduced to each other before the course started. Pupils took part in organised learning activities and accessed the canteen at break times. They travelled to and from the school by bus.

In successfully completing this unit, the pupils would demonstrate the following abilities:

The mainstream school is well equipped with extensive workshops and CAD/CAM facilities. It was felt that these offered the pupils from the special school an opportunity for a different experience and the chance to make good quality products.

Table 4.1

Setting off

1. Collect own belongings.	2. Find way to mini bus.
3. Sit in own seat.	4. Put seat belt on.
5. Travel in an acceptable manner.	6. Request assistance.

At CTC

7. Recognise and greet pupils.	8. Find way to signing-in desk.
9. Wait turn at desk and remain with group.	10. Find way to cloakroom.
11. Use lift.	12. Use stairs.
13. Find way to learning 'base'.	14. Complete/participate in learning activity.

End of lesson/break

15. Find way to canteen.	16. Use vending machine to buy drink.
17. Behave appropriately in the company of peers.	18. Interact positively with peers.

Pupils' work began with a simple clock, which was based on a vacuum-formed plastic body. Each pupil produced the clock body using a range of machinery and were assisted by a small group of mainstream pupils. There was no intention of providing any designing activity at this stage. The former had been produced so that the plastic body would offer a good surface area to personalise the product.

The next stage was to use a computer program to design suitable graphics which would eventually be cut from self-adhesive vinyl on a plotter cutter. A template had been prepared to allow the pupils to size and space the graphics accurately.

The pupils quickly got to grips with this and, apart from the final positioning onto the clock body and assistance with fitting the mechanism, they raced ahead of schedule. It was at this point that the work took an unexpected turn as there was unplanned time to spare. It was decided that a sign should be made for a room at the host school, and this developed into a complete signage system which extended the project into another academic year. The royal blue plotter-cut vinyl signs were stuck onto yellow polystyrene sheet backing plates which the pupils cut on a bandsaw and cleaned carefully. Double-sided tape was applied to each sign so that it could be stuck to a door at the school. Several groups of pupils took part in the work and it was easy to slot a different group into the signage production, as batches of signs were produced: they were printed with phrases such as 'Privacy and Dignity' and with room numbers, and so on.

A year into the partnership and, flushed with success, it was felt that more ambitious work could be undertaken. The pupils investigated some needs and opportunities back at school and returned with requests for number matching. A milling machine was put to use and the results were very professional. A wide range of traditional woodworking machinery, such as bandsaws and sanders, were utilised to complete the task.

While it is clear that the special school benefited considerably from the partnership, it is perhaps not so obvious what it was that the host school gained. Several post-16 pupils worked closely with the special school pupils and gained in confidence and maturity as a result. Indeed, two of them went on to undertake GCE A-Level design and technology major projects aimed at children with special needs. Staff have benefited from working with different pupils who have different needs and the whole college benefited from having special school pupils take part in normal social events.

The messages in this experience are clear. Large secondary schools often have equipment which could benefit special needs schools' pupils. CAD/CAM offers particular benefits in terms of accessibility and quality of results. Mainstream pupils benefit from this type of social interaction and soon become very adept at knowing when to support and when to step back. It is anticipated that this partnership will continue indefinitely.

5 Special designers and makers

Giftedness and learning difficulties

We, as teachers, sometimes think negatively of a learning difficulty, such as dyslexia and autism. It is important for D&T teachers to recognise that some learning difficulties contribute to pupils' D&T capability in quite remarkable ways. Ofsted often reports that pupils with SEND make more progress in D&T than in any other subject. The hands-on, creative problem-solving and open-ended, 3D nature of the topics means that pupils excel in this subject where they might struggle in others. Pupils who are both gifted and learning-disabled (sometimes referred to as 'dual exceptionality') can exhibit remarkable talents and strengths in some areas of D&T, but may also demonstrate significant weaknesses in other areas of the subject.

These pupils' hidden talents and abilities may emerge in D&T or may be stimulated by a classroom teacher who uses a creative approach to learning.

Particular efforts also need to be made to ensure that pupils with disabilities who are talented in D&T are identified and that appropriate provision is

Table 5.1

Possible strengths include	Weaknesses may include
Keen visual memory,	Distractibility and/or disorganisation
Well-developed spatial skills	Difficulties with sequential tasks
Imagination and creativity	Super-sensitivity
Insight	Failure to complete assignments
Good problem-finding and solving skills	Poor 'stickability'
Perfectionism	Perfectionism (when expectations are unrealistic)
Wide variety of interests	Obsession with one particular topic
Comprehension of complex systems	

made to assist them in overcoming difficulties they face and developing their abilities. For example:

- Pupils who are hearing impaired may lack vocabulary and linguistic skills and have difficulty in responding to oral questions or expressing themselves in a way that reflects the complexity of their thoughts.
- A pupil's visual impairment may result in particular problems with work relating to developing their practical skills.
- Pupils who have specific learning difficulties may find it challenging to plan and organise their work, but may have good problem-solving skills.

These pupils are often noticed because of what they *cannot* do, rather than because of their talents, and staff may focus on their 'problems' rather than on their strengths and interests. Since they are bright and often sensitive, they are more acutely aware of their difficulty in learning and if they fail to complete a task, they feel inadequate. However, these youngsters often have high-level interests at home and will tell you about fantastic models or structures they have made. The creative abilities, intellectual strength and passion that they bring to their hobbies are clear indicators of their potential for giftedness. Research has shown that this group of pupils is often rated by teachers as most disruptive at school (see the reference list in chapter nine for reading on this). They are frequently found to be off-task; they may act out, daydream, or complain of headaches and stomach aches; they are easily frustrated and may use their creative abilities to avoid tasks.

Examples of designers and technologists with learning difficulties

There are a number of well-known designers and technologists who are known for having learning difficulties. Far from hampering their success, sometimes their condition has contributed to their talents. Here are some examples of people with significant intellectual disabilities, such as autism, and dyslexia who have a 'fragment of genius' in D&T.

Autism (including Asperger syndrome)

There are some remarkable individuals who have extraordinarily specific abilities while being extremely limited in all others. One example is Steven Wiltshire, who was taught by my colleague for a time at Wood Lane School. Steven has a remarkable talent for drawing that has made him very successful and he is now a professional full-time artist. For more information on Stephen, please see his website: www.stephenwiltshire.co.uk.

He is also musically gifted and has perfect pitch. At school, he was very fond of the practical side of food technology and later signed up for catering college.

Stephen was born in London to West Indian parents in 1974. As a child, Stephen was mute and could not relate to other human beings. Aged three, he was diagnosed as autistic. At Queensmill Special School, teachers noticed that he enjoyed drawing animals, buses and buildings. His drawings were extraordinarily mature and demonstrated a natural gift for perspective and ability to remember and reproduce the detail in each structure.

Stephen is the only artistic autistic savant in the world whose work has been recorded and published since his childhood and he has appeared on numerous TV programmes.

> It has often been said that savants have photographic or eidetic memories, but as I photocopied Stephen's drawings I thought how unlike a Xerox machine he was. His pictures in no sense resembled copies or photographs, something mechanical and impersonal – there were always additions, subtractions, revisions, and, of course, Stephen's unmistakable style. They were images that showed us some of the immensely complex neural processes that are needed to make a visual and graphic image. Stephen's drawings were individual constructions, but could they be seen, in a deeper sense, as creations?
>
> (Dr Oliver Sacks, *An Anthropologist on Mars: Seven Paradoxical Tales*, Picador, 1995)

Scientists in Australia and the USA think they have identified the part of the brain (left arterial temporal lobe) which, if 'switched off', can stimulate artistic genius. They believe that ordinary people may one day be able to 'tap in' to allow them at least a moment of genius. They think that when a specific part of the brain does not work properly, abilities in another area may be 'unlocked' and that the savants have their gifts because of this 'malfunction' of the brain, not in spite of it.

For example, Stephen Wiltshire was taken up in a helicopter over London. Hours later, he produced a detailed and accurate drawing of a four square mile area of the city. The scientists believe this was possible because instead of his brain processing details of information, such as identifying a building or recognising it, he was able to tune in to all the complex mental processes that lie behind that recognition, and copy them.

Dyslexia

Humans communicate through symbols (speech, written words and signs). Our ability to learn to encode and decode these symbols into meaningful ideas cannot be traced to a specific organ, but is wrapped up in the complexities of neurological brain development. The links between us understanding how our brains work and how we should teach reading and writing are strong. Not everyone is going to learn in the same way! It is not clear that dyslexia is *intrinsically* a defect. It may offer an alternative way of seeing, almost providing a 'three dimensional' viewpoint in some instances. Of course, this is not very helpful when a two-dimensional activity like writing is required. Again, it is not always helpful to regard this condition as some kind of 'disease'. Some, perhaps all, dyslexic pupils gain an advantage of some sort by being dyslexic, though what they gain may not always be acknowledged or appreciated by the educational world.

For some designers and technologists dyslexia has provided a different viewpoint and with this has emerged some of our greatest thinkers and inventors.

Two qualities in dyslexics are often outstanding:

- The understanding/awareness/management of 3D form and space;
- The understanding/awareness/management of innovative composition.

Albert Einstein, Thomas Alva Edison, and Leonardo Da Vinci all experienced early learning disabilities, centering on memory difficulties, and limited abilities in reading, retention and recall. To compensate for and overcome leftbrain weaknesses in verbal retention and reading/writing, these great inventors significantly developed their rightbrain strengths of visualisation and detailed patterning, leading to discoveries that have changed the study and nature of physics, provided numerous daytoday electrical applications, and altered the very way we view our world.

Hans Albert Einstein, on his father, Albert Einstein

'He told me that his teachers reported that . . . he was mentally slow, unsociable, and adrift forever in his foolish dreams.'

Thomas Edison

'My teachers say I'm addled . . . my father thought I was stupid, and I almost decided I must be a dunce.'

Tommy Hilfiger, fashion designer

'I performed poorly at school, when I attended, that is, and was perceived as stupid because of my dyslexia. I still have trouble reading. I have to concentrate very hard at going left to right, left to right, otherwise my eye just wanders to the bottom of the page.'

In reference to his being the class clown, he said, *'I didn't want anyone to know that I didn't get it'*.

Marco Pierre White, chef and restaurateur

'Like many people with a handicap, I compensated elsewhere. When I had difficulty with spelling and reading, I concentrated on mathematics and sports. However, back in class, I found traditional teaching methods such as standing up and reading aloud in class pure torture. Dyslexia gave me a different way of looking at things. A compulsion to dissect ideas and concepts from every possible angle has stayed with me.'

Richard Branson, Founder of Virgin Enterprises

Richard didn't breeze through school. It wasn't just a challenge for him, it was a nightmare. His dyslexia embarrassed him as he had to memorise and recite word for word in public.

Frustrated with the rigidity of school rules and regulations, and seeing the energy of pupil activism in the late 60s, he decided to start his own pupil newspaper. It would have articles by ministers of parliament, rock music stars, intellectuals and movie celebrities. It would be a commercial success.

(www.johnshepler.com/articles/branson.html)

William Hewlett, Co-Founder, Hewlett-Packard

Hewlett was hampered by undiagnosed dyslexia, which gave him difficulty with written assignments, but led him to develop exceptional memorisation and logical skills. Hewlett excelled in mathematics and sciences. (www.obits.com/hewlettwilliam.html)

Craig McCaw, telecommunications visionary

'Growing up, I had trouble fitting in', said Mr McCaw in a 1998 interview. 'As a dyslexic, I don't think like other people, so I didn't fit very well in a clique.'

Mr McCaw is shy and unassuming, visibly uncomfortable during his rare public speaking engagements. The narrative of his ideas is disjointed, his point only becomes clear when his trains of thought collide in an unpredicted conclusion. He is famously well-known for blowing the punch-lines of jokes.

It is important to understand how his dyslexia has influenced significantly his entrepreneurial vision. Mr. McCaw credits his ability to see circumstances from unique perspectives; to see, for example, the potential of cellular communications, an insight that seems obvious now but that was uncommon in its day. 'Dyslexia forced me to be quite conceptual, because I'm not very good at details', he said at his 1997 induction into the Academy of Achievement. 'And because I'm not good at details, I tend to be rather spatial in my thinking – oriented to things in general terms, rather than the specific. That allows you to step back and take in the big picture. I feel blessed about that.'

Charles Schwab, investor

Along the way, I've frustrated some of my associates because I could see the end zone of a particular thing quicker than they could, so I was moving ahead to conclusions. I go straight from step A to Z, and say: 'This is the outcome. I can see it'.

Dr Sally Shaywitz, director of the Learning Disorders Unit at Yale University says, 'Mr Schwab's ability to see solutions that others cannot is typical of dyslexics. What distinguishes them is that they really think outside of the box.' Dyslexics, she said, 'often have a variety of qualities, including resilience, adaptability and the ability to formulate original insight'.

Thomas G West and Daniel J Sandin

Both Thomas and Daniel have dyslexia and are incredibly creative. Thomas West says that dyslexic people are often highly visual, able to quickly process and integrate high-quality visual and spatial information. He notes also that, 'Society is shifting, with far more call for that sort of skill than for more mechanical tasks like reading and writing.'

People with dyslexia seem to problem-solve in unusual ways, perhaps working from the inside out or from the back to the front. Sandin, who says he still cannot spell or do arithmetic, talks about developing the CAVE virtual reality system, which uses rear projection screens instead of a headset and knows where you are by generating your position in the room. It is a visual simulator completely matched to the human visual perceptual capability, he says. Thomas West says that, 'The dyslexic person may well be at the fore as the technological revolution continues, with their ability to process information and data and depict visually creating a whole new literacy.'

Anne Cunningham, Education Projects Officer, The Lighthouse

When people talk about dyslexia, it's usually in terms of failure and not being able to do things. But there is much that I, and other dyslexics, can do that other people struggle with. For instance, we're often good with 3D and visual and spatial comprehension, and as we are approaching a problem differently, dyslexia can help with creative problem-solving, which is why people working in teams like having us around.

(Anne Cunningham, Education Projects Officer, The Lighthouse)

Curricular Needs

Although each pupil is unique, they all require an environment that will develop their particular gifts; this will necessitate emotional support and provisions to address their learning disability. Four general guidelines can help to meet the needs of these pupils.

Nurture their gift

In the past, teachers have focused on addressing the learning difficulties first, but it is important not to focus on the weaknesses as this can result in poor self-esteem, a lack of motivation and even depression and stress. Instead, we should develop their strengths, interests, and superior intellectual capacities with a stimulating classroom environment and enrichment activities designed to minimise weaknesses and highlight creative thinking and innovation.

Provide a nurturing environment that values individual differences

Success in the real world depends on skills or knowledge in other areas besides reading and writing. D&T provides an opportunity for pupils to contribute and feel valued, and to better understand their own needs and wants. A nurturing D&T classroom is one that develops individual pupils' potential and values and respects individual differences. Pupils are offered different ways of approaching a design brief, which plays to the strengths of their learning style. They can respond to researching and presenting in a way that works for them. Remember, these pupils do not want the curriculum to be less challenging or demanding. Rather, they need alternative ways to work. They are encouraged to work collaboratively in teams and to support each other. Pupils are rewarded for what they do well.

Encourage compensation strategies

The following list outlines suggestions for providing compensation techniques to assist pupils:

- Find sources of information that are appropriate for pupils who may have difficulty reading, such as: videos, visual websites, games, visits, interviews, photographs and practical work.

- Provide organisers to help pupils to receive and communicate information. Teach pupils who have difficulty transferring ideas to a sequential format on paper to use brainstorming and webbing to generate outlines and organise written work. Provide management plans in which tasks are listed sequentially with target dates for completion. Finally, provide a structure or visual format to guide the finished product. A sketch of portfolio contents pages will enable these pupils to produce a well-organised product.
- Use ICT to promote productivity. It is efficient to organise and access information, increase accuracy and enhance the visual quality of the finished product. It allows pupils with learning disabilities to submit work of which they can feel proud.
- Offer a variety of options for the communication of ideas, not just written – recorded, digital photo stories, annotated sketchbooks, comic strip etc.
- Help pupils who have problems with short-term memory to develop strategies for remembering. The use of mnemonics, especially those created by pupils themselves, is one effective strategy to enhance memory.

Encourage awareness of individual strengths and weaknesses

Pupils need to understand their abilities, strengths and weaknesses so that they can make intelligent choices about their future. Mentoring by adults (for example existing designers) who are gifted and learning disabled will help them believe that such individuals can succeed.

Professional development activity

See the accompanying website (www.routledge.com/9781138714922) for a selection of 'pupil pen portraits', some of which you may recognise amongst your own classes. The 'strategy cards' provided, suggest starting points for staff to consider in planning appropriate D&T assignments for these pupils.

6 Monitoring and assessment

Planning and reviewing progress

> Pupils need to be assured that their successes will be recognised and rewarded and that, when they have difficulty accessing learning, the necessary support will be available. Teachers need to know when things are going well and when they are not, so that they can change their teaching approaches as appropriate.'
>
> *(SEN: Training materials for the foundation subjects*, Introduction, DfES 2003)

Identifying starting points

D&T teachers use a variety of assessment techniques to identify the needs of individual pupils, for example, a standard practical task, observation checklists for development of practical skills, or an opportunity for a pupil to explain or communicate a design idea to the group. It would be rare for the teacher to use a long project as an assessment task for pupils with special educational needs. They may use part of the main project or a shorter activity that is observed by the teacher, such as following a sequence of instructions to assemble a simple product, or using a chart template and word bank to evaluate and test a set of existing products. These are important to identify starting points from which progress can be measured.

These opportunities to identify a pupil's progress help the teacher to gather information about existing knowledge, skills and understanding, as well as strengths and needs, personal interests and how the pupil prefers to learn. The priorities can then be recorded as targets on any individual plan devised for the learner. For pupils with significant needs, an education, health and care (EHC) plan will be in place, helping to determine learning priorities and targets. This will also provide valuable indicators regarding how much support the pupil needs in order to access and complete tasks, which will help with planning resources and support.

For all pupils with special educational needs, whether or not they have an EHC plan, the continuous 'Assess, Plan, Do, Review' cycle, is key to effective teaching and learning.

(See Appendix 6.1 Differentiation and progression steps.)

Working from the EHC plan

The EHC plan, or any type of individual pupil plan, should set out the 'what', 'how' and 'how often' of special provision. In other words, it should identify those activities that are additional and different from those provided for all pupils through the differentiated curriculum. Setting out the steps required to help pupils achieve identified outcomes, these plans should describe short-term targets and strategies, with clearly specified success criteria. It is often helpful to use phrases such as, 'By the end of term, Barrie will be able to . . .'. Setting too many targets at one time is not appropriate: it is best to prioritise up to three or four key objectives.

Such plans are working documents and as such they should be in regular use, annotated and amended as necessary and shared amongst all staff involved with the pupil, as well as with the pupil himself/ herself of course – and parents.

Individual plans should:

- detail provisions additional to, or different from, those generally available to all pupils;
- inform effective planning;
- help pupils monitor their own progress;
- result in the achievement of specific goals for pupils and raise attainment.

Writing S.M.A.R.T. D&T targets

Appendix 6.2 is a reminder of the pertinent points to remember when suggesting and agreeing targets with pupils. A 'method' section aligned to targets should be used to detail the circumstances or contexts in which the targets can be addressed – activities, resources, equipment, staff roles, pupil groupings, rewards, etc.

Opportunities and activities at Key Stage 3

Much of the D&T Key Stage 3 is relevant to pupils with learning difficulties. With modification, it can provide stimulating and challenging learning opportunities.

The focus of teaching D&T at Key Stage 3 may be on giving pupils opportunities to:

- suggest outline plans for designing and making
- communicate design proposals
- select and use tools, equipment and processes, including CAD/CAM
- explore the properties of a range of contrasting materials
- analyse products and judge the quality of others people's products

Given these opportunities at Key Stage 3

all pupils with learning difficulties (including those with the most profound disabilities):

- make choices about a product or aspects of its design
- observe, explore and experience a range of materials and tools

most pupils with learning difficulties (including those with severe difficulties in learning) who will develop further skills, knowledge and understanding in most aspects of the subject:

- select and use a variety of tools, equipment and processes
- communicate design proposals in a variety of ways
- test and refine their ideas against a simple specification

a few pupils with learning difficulties who will develop further aspects of knowledge, skills and understanding in the subject:

- identify and solve their own design problems
- take into account the views of intended users and other interested groups

There are some parts of the PoS that may be too demanding, for example, at Key Stage 3:

- generating design proposals
- prioritising actions and reconciling decisions as a project develops
- considering the chemical and physical qualities of materials
- understanding systems and control or structures

Monitoring pupils' success against medium-term and short-term objectives

It is helpful to consider both medium-term and short-term objectives for the individual.

- *Medium-term learning objectives* should be set out which provide priorities for learning. These may be written as key skill objectives, for example communication and independent learning, or in terms of the pupil's priority needs. A good plan will include success criteria related to these. The objectives and success criteria will be reflected in the planning for the unit of work or design and make assignment or project.
- *Short-term learning objectives* should be clear for individual lessons or tasks, such as focused practical tasks or product evaluation activities. For some pupils, objectives should be set for separate episodes of learning within the lesson. These objectives need recognisable success criteria.

Example – Planning and reviewing progress against medium-term learning objectives and success criteria

Puppets project

A class teacher is working with some Year 7 pupils on a puppets project.

Design and make assignment

Puppets are a good way to help children learn about issues such as crime, avoiding strangers, road safety, keeping teeth clean, healthy eating, not playing with fire; and a range of social skills such as good sharing, dealing with bullying and coping with unfamiliar situations. Pupils can be tasked with producing a puppet and writing a story to teach primary school children something useful.

The teacher writes an assessment guide sheet (see opposite):

This assessment sheet clearly identifies:

- assessment focuses for the project;
- success criteria for the project (your learning outcomes);
- success indicators at different levels (basic, intermediate and advanced statements).

Effective assessment and record keeping can be supported by:

- specifying time for observation in a unit of work;
- targeting specific pupils for observation and recording particular lessons, ensuring that all learners are assessed in all subjects over time;

This assessment will focus on:

Designing: with the use of colour and texture in a wide range of possible materials.
with simple mechanics to produce an animated puppet.
Making : with quite low levels to high levels of finish.

Your Learning Outcomes

- You will learn how to emphasise the special features that give someone or something their own particular character.
- You will be responsible for a choice of materials. This will allow you to be creative and imaginative.
- This challenge will add to your knowledge of how puppets are used around the world and through the ages.
- You will be developing your imagination as you bring your character to life

Advanced Designing

- makes the fullest use of quality drawing and models
- demonstrates competence in visualisation techniques
- contains designs which are derived from a range of sources
- ideas portray a character, with vitality.
- evaluation shows understanding of 'character'.

Advanced Making

- made to a standard that allows it to work safely, accurately and visually

Intermediate Designing

- drawings and / or models to an adequate standard
- competent use of either colours or materials in creating a character.
- a limited number of alternative designs
- some range of design sources.
- final design proposal is shown with reasonable accuracy.

Intermediate Making

- made in a planned manner with care and accuracy
- evaluated to show where faults have occurred and simple changes that could be made to improve some aspects of the finished product.
- works adequately functionally / visually

Basic Designing

- labelled sketches
- reasonable planning to produce the puppet character
- some understanding of the way that colours or materials have helped to create the puppet character.
- a final design proposal.

Basic Making

- relates to the original design intention
- is made to a functional level of accuracy or appropriate visual quality
- is evaluated to identify simple changes that could be made to improve an aspect of the finished product.

Figure 6.1 Assessment sheet: Puppets project

- giving responsibility for observation and record keeping to named members of staff in specified lessons;
- involving pupils in assessment and recording processes.

Sharing objectives with the staff team

Before the lesson begins discuss the following points with the team:

- how to check for signs of success;
- how information about the pupil's understanding and misunderstanding will be communicated during the lesson.

The classroom team can then:

- help pupils to recognise their own success in each objective;
- ensure that success, where possible, is identified, even if it is shown by a single remark or other response;
- provide feedback if there is still a lack of understanding so that a different learning approach can be tried.

The importance of sharing objectives and success criteria with pupils

Pupils should be clear about what they are expected to learn and what counts as success. Research by the Goldsmith's College Technology Education Research Unit (TERU) shows that pupils perform for individual teachers in relation to what they think is expected of them. It is important that pupils receive an accurate message from staff members as to what is expected of them. It is even more important that they begin to negotiate their part in defining and achieving these objectives. All adults working with the pupils need to be clear about what is expected of them in a lesson and what counts as success for them. Pupils who have a positive image of themselves in D&T perform to that image. Assessment and feedback reinforces a pupil's self-image and it is important that their experiences of objectives and success criteria are realistic and positive.

(A useful prompt sheet is provided in Appendix 6.3. There are also stems for writing objectives in Appendix 6.4 and a sample lesson plan in Appendix 6.5.)

How you can share objectives

In the puppets example you can see how the teacher shares with the pupils – 'Your learning outcomes' – and includes it in the project booklet introducing the topic. This clearly sets out at the start of a sequence of lessons what will be expected from the pupils. It identifies clear targets for them. (Also see figures 6.2 and 6.3 for examples of how a teacher uses WALT (We are learning to) and WILF (What I am looking for.))

Year 7: Water containers

WALT – We are learning to

1. Understand why water is important

2. Explore and refine strategies for storing water

3.

WILF – What I am looking for

1. Three reasons why water is important

2. A range of ideas for water containers

3.

Starter
How much water do you use in a day? (Ask the pupils to stand next to the measure that displays how much they drink in a day.)

Activity

Introduction

Introduce the importance of drinking water rather than carbonated drinks. Show them an amount of water ie 330ml or 440ml. Possible link with Numeracy looking at measurement and % of a litre.

Episode of Learning 1 **Explore liquid containers**

Look at different kinds of ways of containing liquid. Examples of cartons/bottles required to look at.

Use PMI (Plus Minus Interesting) for each of the different types.

Episode of Learning 2 **Explore ideas from nature** about storing and hydrating (for example, animals, waterfalls, glaciers)

Think of imaginative and attractive ways of carrying and storing the amount of liquid a pupil needs to use and re-fill each lesson.

Episode of Learning 3 Pick an idea and **use SCAMPER** to develop the idea.

Plenary
Think of a slogan to make people drink more water by using the letters WATER

Figure 6.2 Water containers

Year 7: Reflectors (for cycling/walking in the dark)

WALT – We are learning to

1. identify key criteria points for making a reflector

2. model ideas quickly and evaluate their effectiveness

3.

WILF – What I am looking for

1. a range of criteria to be generated

2. imaginative models

3. clear discussion of the merits of the ideas.

Starter
Game - http://talesoftheroad.direct.gov.uk/be-bright.php
Brainstorm key points.

Activities

Introduction

Explain that they are going to design and make a personalised reflector to be attached to their rucksacks or coats to make them stand out in the dark.

Episode of learning 1 **Quick modelling**: Take a sheet of A4 paper and cut vertically into 4 strips. Allow 5 minutes to make as many models of a reflector that fits the brief.

Episode of learning 2 **Evaluation and discussion** Gather all models together and conduct a quick evaluation leading to Key words – criteria points for the design. Devise a list of important factors – shape and size, materials, reflecting material, personalisation. Keep a copy for reference throughout the project.

Episode of learning 3 **Explore products** already on the market – pictures for discussion or handling collection – key questioning about material, style, size and shape

Plenary
Looking at the picture of the child where do you think would be the best place for reflectors?

Figure 6.3 Reflectors

In addition, the teacher uses a range of strategies for sharing objectives with pupils, making sure they are displayed so that teaching and support assistants can refer to them during the lesson and 'mini plenaries'. Remember to:

- Reinforce objectives at the beginning of the lesson and, where appropriate, during the lesson in a language that pupils can understand.
- Agree the objectives with the pupils as an important aspect of the project: 'What will you learn from making a puppet?', 'What did you learn when we did . . .?'
- Display the objectives in the classroom in a variety of ways – in their written form, with captioned photo cards, and by attaching them to a working display of puppets: 'How could you test this puppet? . . . Pull the strings? . . . Ask a young person to use it? . . . Try out a different colour?'
- Link the objectives back to previous relevant work: 'Do you remember when we tested our storybooks to see if young people liked them? How will we test our puppets?'
- Use these objectives as a basis for questioning and feedback in plenaries: 'What is the character of this puppet? What kind of puppet would be sad, scary, kind?
- Repeat the objectives during group work (and ensure support assistants do so).
- Give the pupils timing prompts – 'You have 10 minutes to . . .' – to help them meet the objectives and maintain focus and engagement.
- Review learning with individuals against the objectives at the end of the lesson.
- Negotiate new targets with pupils at the end of the task/project to inform planning for the next design and make assignment.

Helping pupils know and recognise the standards they are aiming for

During a design and make assignment, one way to help pupils to see what is expected of them is to show them examples of work from previous lessons or other places. This has to be carefully managed, particularly where pupils lack confidence and think 'I can't do this; I'm no good at this'. Pupils should be shown carefully chosen examples of what others with similar abilities have made, rather than work that is beyond them. The examples should be challenging and exciting without being de-motivating. It can help pupils to focus on what is ahead and what they are about to design and make. This is easier to approach if the pupils have had a chance to negotiate what they are designing and making in the first place.

It is important to show pupils how they are going to improve, the journey of how someone increases their skills and understanding through focused practical

My Practical Assessment

Tick the column you think You achieved today

	Working towards	Almost there	I achieved if
Me the Chef			
I had good personal hygiene			
I selected the correct equipment			
I weighed and measured accurately			
I used a knife or peeler correctly			
I followed the recipe			
I prepared the ingredients correctly			
The Cooker			
I used the hob safely			
I used the oven safely			
How YOU worked today			
I worked safely			
I worked independently			
I was a peer genius – I helped others			
My personal target for next time			
Next time I am aiming to			

Product

Name _____

Date _____

Now write or draw today's ingredients in its food group section of the Eatwell Plate

Bread, rice, potatoes, pasta

Fruit and vegetables

Milk and dairy foods

Meat, fish, eggs, beans

Foods and drinks high in fat and/or sugar

I think today's recipes were **healthy – unhealthy** (circle your choice)

I think this because _____

I could make this product healthier (more nutritious) by _____

Figure 6.4 Practical assessment

tasks and product evaluation, how they practice and then work further on their project. Pupils should be reassured that they are going to be taught how to do things and supported, so that they know that they will be successful.

It can also be helpful to show how some leading designers do not succeed first time around, so that they get the feeling that 'we are all learning'.

Involving pupils in peer and self-assessment

As with all pupils, where possible, those with learning difficulties should be involved in monitoring their own progress. Pupils need to be encouraged to reflect overtly and in a detailed way on their achievements. Give pupils the opportunity to communicate about what they have learned and what they have found difficult. Developing confidence and raising self-esteem are integral parts of this process. Visual records contribute to pupils' sense of achievement – this is reflected well in the pride that pupils take in recording their achievements in their achievement files. Access to the record-keeping system is important if pupils are to be enabled to refer to their targets.

For example, during the design and make project, perhaps towards the end, plan some time to review the activities. Use a visual storyboard of the project that pupils are undertaking and, together with the pupils, review their progress towards their D&T targets ('What can you do now that you couldn't do before?'), discuss task 'likes and dislikes' and record comments in an individual planner. Use these to set some targets for the next lessons and record this: 'In the next lesson I will . . .'.

Encouraging reflection

- Having asked pupils to reflect on their learning experience, allow sufficient wait time for responses.
- Make use of careful questioning helps pupils to remember and elaborate upon their responses.
- Model how you reflect, for example how you consider alternatives and their implications (think out loud).
- Encourage pupils to compare this project to previous work. This places heavy demands on memory, where pupils are expected to recall activities and their own performance. Use concrete materials, as well as visual images in the form of photographs, video or computer-generated images, to assist them e.g. using digital pictures and items from the portfolio of a previous project.
- Encourage pupils to share opinions with others and discuss how to improve, showing their work to the group and giving and receiving feedback. Social interaction can be supported through the careful selection of resources and pupil grouping.

| | 😊 | 😞 | Evidence |
	YES	NO	(add an image or link here)
I came up with design ideas by...			
talking			
writing			
drawing			
modelling			
making			
To help make it, I used a ...			
drawing			
plan, recipe or instruction sheet			
time plan			
list of tools and equipment			
At the end, I...			
was pleased with it			
looked for ways to improve it			
tested it out on other people			

In my next project I need to work on

1 _____

2 _____

3 _____

Figure 6.5 Reviewing my designing and making

Providing effective feedback

Teaching staff and support assistants can give feedback to pupils continuously through the lesson. Oral feedback is particularly helpful in D&T in addition to (or instead of) written feedback. Ensure that feedback is constructive and focuses on what the pupil has done well, what they need to do to improve and how to do it. For example, if the pupil is in the middle of making a bag, focus on what processes they have done well (pinning, cutting, sewing, accuracy, neatness) and what they need to do next to improve (concentrating carefully when using the sewing machine and controlling it slowly to produce a straight seam).

Remind the pupil of their targets before the learning starts and negotiate the next steps for learning. Use short-term targets to help pupils to progress. For example, use the pupil's own designs and products, at differing stages of completion, to ask the question, 'What next?' during a plenary.

Promoting pupil confidence

Build pupils' confidence and self-esteem by:

*   Using small steps to enable pupils to see their own progress. For example, in reviewing existing products, pupils can be rewarded for looking at each individual product rather than the set; perhaps aspects such as colour, material used and shape.Planning projects so that pupils can use previously gained skills in new situations (for example, mixing dough).
*   Creating a secure, supportive environment for pupils to explain their ideas and their thinking.

Recognising progress and assessing without levels

Assessment needs to be viewed in its widest sense in the context of D&T – it is not just about marking. Making regular and frequent checks on pupils' acquisition of knowledge, understanding and skills is essential in order to judge their pace and progression. What is required is the compilation and recording of a range of evidence on which teacher judgements can be based. Such evidence can be gained through observation of pupils while they are working on tasks, combined with folder work, products and information gained while talking to the pupil. Pupils with learning difficulties often produce very little written or supporting evidence for their work. They find it difficult to record their thoughts and design decisions. If the teacher looks only at the folder and the end product made by the pupil they are often faced with evidence that reveals little of the actual quality of design thinking or depth of knowledge held by the pupil.

For all pupils, including those with learning difficulties, progress is about change and development. For most pupils with learning difficulties, achievements can be predicted and planned for, and progress can be demonstrated in terms of increased knowledge, skills and understanding. Some may follow the same developmental pattern as their fellow pupils, but not necessarily at the same age or rate. Progress may be made in some areas of the curriculum but not in others. For some pupils, progress may be difficult to predict or idiosyncratic and may only be demonstrated in a certain environment with a specific person or materials.

Progress may be recognised when pupils with learning difficulties:

- develop ways to communicate, from the use of concrete actions (pointing to materials and tools in front of them they want to use), to the abstract (using pictures, symbols, print, signs, ICT and the spoken word to ask for materials and tools);
- change responses to social interactions, from passive cooperation toward active participation with individuals and groups;
- need less support, for example from a helper, technology or individualised equipment, in carrying out tasks;
- develop a wider range of responses to experiences, even if no new skills and knowledge are gained, for example when evaluating products;
- increase their knowledge and understanding about D&T;
- maintain, practise, or combine skills over time and in different situations;
- become more flexible in learning styles and environments, reducing the need to present activities in consistent and personalised ways;
- behave in a way that does not inhibit learning frequently;
- cope with things that they were unable to previously, for example with change or new situations, with trying out new designs or modifying their plans;
- choose not to participate or to respond, for example, choosing not to make a particular recipe.

Now that there are no national level descriptions, D&T teachers are able to devise progress statements against what is taught in their scheme of learning. This means that individual schools will adopt their own set of expectations for each year group against their unique teaching plan. However, it is useful to check if these are in line with national expectations by:

- working with colleagues in other schools and exchanging example assessment descriptions;using any available national frameworks that set out what professional bodies consider appropriate (for example DATA, or Public Health England Food competences);
- reviewing past QCA P Level or performance descriptions as, whilst these have not been updated (nor are they a requirement), they do provide a valuable starting point for writing progression statements. These performance

descriptions are not intended as a tool to measure hierarchical and linear progress in a mechanical fashion from encounter to attainment, but may give teachers a greater understanding of how pupils move through a learning process. Teachers may wish to use this framework to develop their own assessment tools so they take into account the differing needs of pupils such as those described below. (The performance descriptions can be found in Appendix 6.6.)

Jack is a Year 7 boy who has severe learning difficulties and epilepsy as a result of serious illness suffered when he was a young child. In D&T, he finds it difficult to record designs and plan his own work and needs a significant amount of adult encouragement to participate actively in lessons. Jack can communicate using three or four word phrases; he is just learning to copy

Amanda was working on a puppets/Joseph's coat project in D&T textiles

Amanda was able to select a garment she wanted to wear. She was able to select from two fabrics for dressing a doll for a party context. She was able to draw a repeating pattern by being given one crayon at a time, i.e. red then blue, then red, etc. Amanda could select dye colour for batik patterns and was able to perform threading action with laces. With help, she could use a lace to join together the fabric pieces she had selected. She was able to select buttons and sequins but needed help in attaching them to her garment. Her comment on her finished garment was 'nice'.

Lewis was working on the puppets project in D&T textiles

Lewis decided his plate puppet would be of a man. He discussed his ideas and considered people he knew as well as characters in books for ideas. His ideas developed during discussions and while looking at the materials available to him. He was able to communicate his ideas by producing a labelled diagram on a pro forma design sheet.

During the making phase, Lewis demonstrated that he was able to select appropriate tools and materials, and mark out, cut and shape the materials. He was also able to assemble the components.

Evaluation was ongoing, during which Lewis was able to talk about his ideas. On completion of the product, Lewis was able to say what he liked and disliked, but was not able to identify ways in which he could improve his work in the future.

Figure 6.6 The teacher kept copies of some work as evidence of Jack's progress

Figure 6.7 Jack was working on
a Snack Bar project
in D&T Food (Cooking
and Nutrition)

write and he is at an early reading stage. In this project, Jack had to try to give his opinion about different snack bars and make choices to develop a healthy recipe from given ingredients.

Jack was not keen to taste or give opinions about the commercially-produced snack bars, but with encouragement he would indicate which of them he had tried. He worked in a small group to experience following recipes (with symbols) with staff help and, with much encouragement, made his own choice between cake-based and biscuit-based snack bars. He practised putting foods into the three traffic light groups.

Jack chose his additions to his plain cake-based snack bar recipe (with muesli) and made it with adult assistance. Jack could copy the actions of the assistant and said his choice was 'good' when asked.

He enjoyed the experiments with different wrapping materials and chose a suitable material to wrap his bar. Jack was helped to write his choices on his worksheets, to which symbols were added to aid his understanding.

Not all pupils will make progress, however. Staff will recognise that, because of their learning difficulties, some pupils may reach a plateau in their achievements, or regress. This is usually temporary, but sometimes can be lengthy or permanent. In such cases, pupils' recorded attainments, or achievements previously predicted by staff, may decline. A slowing of the rate of regression, shown by skills or capabilities being maintained or reactivated, is then viewed as a form of progress. You may find the following framework useful in describing small steps of progression.

A framework for recognising attainment

Encounter

Pupils are present during an experience or activity without any obvious learning outcome, although for some pupils, for example, those who withhold their attention or their presence from many situations, their willingness to tolerate a shared activity may, in itself, be significant.

Awareness

Pupils appear to show awareness that something has happened and notice, fleetingly focus on or attend to an object, event or person, for example, by briefly interrupting a pattern of self-absorbed movement or vocalisation.

(Continued)

Attention and response

Pupils attend and begin to respond, often not consistently, to what is happening, for example, by showing signs of surprise, enjoyment, frustration or dissatisfaction, demonstrating the beginning of an ability to distinguish between different people, objects, events and places.

Engagement

Pupils show more consistent attention to, and can tell the difference between, specific events in their surroundings, for example, by focused looking or listening; turning to locate objects, events or people; following moving objects and events through movements of their eyes, head or other body parts.

Participation

Pupils engage in sharing, taking turns and the anticipation of familiar sequences of events, for example, by smiling, vocalising or showing other signs of excitement, although these responses may be supported by staff or other pupils.

Involvement

Pupils actively strive to reach out, join in or comment in some way on the activity itself or on the actions or responses of other pupils, for example, by making exploratory hand and arm movements, seeking eye contact with staff or other pupils, or by speaking, signing or gesturing.

Gaining skills

Pupils gain, strengthen or make general use of their skills, and understanding of knowledge and concepts or understanding that relates to their experience of the curriculum, for example, they can recognise the features of an object and understand its relevance, significance and use.

(See Appendix 6.1 and the web resources for 'Differentiation and progression steps' charts related to the projects featured in this book.)

What records do you need to keep?

Keeping up-to-date records can help staff with planning future lessons and tasks. Records can demonstrate pupils' knowledge and their practical abilities, show what teaching methods have worked well with them and what affects the pupil's performance, such as health or a change in home circumstances.

For pupils with learning difficulties, records of experiences, progress and achievements in relation to targets in their EHC and curriculum plans should focus on significant responses or ways of learning. A system should be flexible enough to include unexpected or unusual responses, however these occur. The needs of individual pupils may determine the type of record, and it may be necessary to draw up individual standards.

It is up to staff to decide the kinds of records they keep. Their decision will be based on how useful they and other staff find the records. Records may include:

- extracts from curriculum plans (as records of experience);
- comments about pupil responses;
- annotated samples of work;
- photographs, or tape or video sequences;
- pupil self-assessment and peer recordings;
- a pupil's record of achievement or progress file;
- assessments related to external accreditation.

Regular monitoring and recording of pupils' responses and progress identifies areas where pupils are making steady progress and where progress is not being maintained. The responses of some pupils may change from lesson to lesson and topic to topic and may be dependent on factors such as:

- preferences for certain helpers and support assistants;
- grouping or positioning with certain pupils;
- different environments;
- the time of day;
- access to their favourite tools/equipment;
- particular sorts of sensory experience, for example tasting, mixing dough and touching products;
- focus material being used, for example plastic, food, metal;
- emerging talents in particular aspects of designing and making.

Accreditation at Key Stage 4

Inspectors found instances, in around 10 of the schools visited, where students with special educational needs and/or disabilities struggled to explain what they knew in response to theory-based written examinations. In these cases, schools were not working closely enough with examination boards to find alternative methods to enable students to convey what they know and can do.

(Ofsted, *Meeting technological challenges? Design and technology in schools 2007–10*)

It is important that teachers work alongside awarding bodies to investigate accreditation that may be appropriate for the learner's needs, and will provide a nationally-recognised certificate for their achievements.

There has been significant change in the GCSE system over recent years, and standards have been set by OFQUAL to make the new courses more challenging, and are increasingly examination paper assessed. This does not suit all learners.

There are some new qualifications in the framework, such as Vocational Certificates (VCerts) and Technical Awards. These are set in the context of vocational learning, and may suit some students better than others, although they still have a written examination requirement. If your school does not require you to choose a course for the performance table measures, then there may be a greater variety of smaller unit-based or awards-based certificates to offer recognition to pupils. For many learners with SEND, a 2-year course, or 120 hours, seems a long time and a large qualification, so accreditation that recognises smaller steps is helpful.

The Unit Award Scheme (UAS) provides a possible alternative to recognise the achievements of learners working at 'Entry Level' (Entry Level Certificates or ELCs have been discontinued for many subjects other than maths, English and science). UAS provides simple and effective recognition for learners achieving short units of accessible learning, and provides a real way to motivate, engage and encourage learners. There are suites of units available which are linked to the old ELCs, which continue to be available for use on a unit-by-unit basis through the UAS.

In the case of D&T, teachers can choose from over 100 Units. These unit awards range from accrediting the achievement of a single skill (such as using a saw, using simple circuits and a Raspberry Pi), and awards for following correct procedures (such as working safely in a sewing room, or safe handling of a pair of scissors), to fuller structured design, make and evaluate projects (such as designing and making a clock or a toy). Some of the unit awards also acknowledge where the task has been completed with assistance, so that there is something for a range of pre-Level 2 GCSE or Technical Award abilities from Pre-Entry Level, Entry Level and Level 1.

Things to consider when choosing KS4 courses:

- Does your school require you to choose a course that is included in the performance table measures?
- How realistic is the content and level of expectation for your learners? How well does it provide for their future needs and provide progression for them?
- What is the nature of the assessments? What percentage of the grade is given for non-examination assessment? Is there any teacher assessment?
- Does the style of examination and non-examination assessment suit your learners?
- How many hours is the course? Is it possible to segment the course into smaller accreditation units if they do not complete the full course?
- Are you able to deliver the course over an extended period of time?
- What special provision is made by the awarding body for those with SEND?
- Is it possible to take the written question paper electronically and/or submit NEA in an electronic format?
- Is it possible to adapt the course materials and assessment so that multiple accreditations are possible in the same class with good planning?

Agreeing special arrangements with the awarding body/exam board

The extent of intervention by support assistants or teachers allowed in practical exam work tends to vary from one board to another. Decide in advance how much support your student is likely to need in a non-examination assessment (particularly the practical elements), and the written examination, and then approach the exam board subject officer to agree special arrangements.

- Generally candidates have to prove that they can do the task or be able to direct their support with exact instructions to do it on their behalf. The candidate is then showing that they know the skill required.
- The teacher may be allowed to take photos or a video to show that a pupil is able to do the stages of a practical task.
- In written work, diagrams may be completed by staff if the candidate has directed the staff member exactly, and a note is made of the direction given and help received.

Summary

- Be clear about the learning outcomes expected from each activity. Assessment is easier if pupils know what evidence they must present and teachers know what they are looking for.

- Try to use as many different assessment methods as possible, using written, drawn, spoken and practical evidence. Some evidence of learning can be written or pictorial, but other evidence arises while pupils are discussing or answering questions. Talk to the pupil and question them as this may reveal a more thorough understanding than their written work suggests. Annotations on pupils' work and notes in the teacher's mark book are an effective way of recording this evidence. If you (or a TA) write the results of the discussion on the pupil's work, this will provide a permanent record and give the pupil a model to help them to know what to write next time. Pupil diaries and review sheets are also useful.

- Assessment should always be positive, acknowledging what has been achieved against set targets rather than making comparisons with the achievements of others. Effective assessment means ensuring that pupils are rewarded for what they know, understand and can do.

- Assessment should be used to give pupils feedback about their work and their learning and to encourage them to assess their own progress, with both teachers and other pupils reviewing what has been achieved and thinking about next steps.

- Recognise and report as many different aspects of the pupil's progress as possible.

- Sensitive observation and diagnostic tests will help to pinpoint areas of weakness in a subject which may have been missed by the pupil and which requires extra attention. They also help to identify strengths on which the teacher can build.

7 Managing support

The ultimate responsibility for a pupil's access to the D&T curriculum is that of the classroom teacher. Support staff facilitate the delivery of an appropriately differentiated curriculum under the direction of the D&T teacher. (Appendix 7.1 gives advice on how to make the most of your support staff.)

In a recent Ofsted report, the valuable role of support staff is highlighted and praised, 'The teaching assistants seen generally made well-judged interventions to guide pupils, particularly those with special educational needs and/or disabilities, but they did not stifle the pupils' independence and creativity by doing pupils' work and thinking for them'(Ofsted, *Meeting technological challenges? Design and technology in schools 2007–10*)

D&T teachers and support staff working together

When working with classroom support staff on a regular basis, the D&T teacher should:

- Give the support staff clear instructions for specific tasks.
- Provide guidance to support staff on how to successfully carry out tasks and maximise the independence of the pupils

Both teachers and classroom support staff should share the following resources:

- the school's policy for D&T;
- the correct way of using tools and equipment;
- health and safety guidelines;
- where to find materials, tools and equipment;
- where to find other resources for D&T;
- practical classroom routines.

The planning and monitoring grids provided on this book's website can help you to plan and record support. There is a weekly planning grid, daily planning grid, weekly support notes and individual support records.

Tips for the teacher on working with support staff

- Direct the support staff and give them clear instructions.
- Explain the dos and don'ts of how you work right from the start.
- Find out about their particular interests and areas of expertise (such as food, textiles, sketching or ICT).
- Be clear about the skills and knowledge they may need to develop and how you can support them to gain these.
- Ask them if they have any concerns and be clear and succinct in allaying them.

Tips for the support staff on working with the teacher

- Find out what is expected of you.
- If you are not sure of something, ask for clarification.
- Don't intrude or take over – remember you are working under the teacher's direction.
- Arrive to classes early to allow time to speak to the teacher and organise resources.
- Be discreet – information about individual children should be treated as confidential.
- Give feedback to the teacher that you believe will be useful.
- Be open and prepared to learn new skills.

Supporting designing and making in D&T

Lessons in D&T will likely involve a support assistant in:

- showing children how to do something;
- talking to children while they are working;
- supporting practical work by the pupils.

Sometimes, they are also asked to supervise group work, organise and maintain resources and create displays to support learning.

Showing children how to do something

One frequent task undertaken by support staff will be demonstrating how to do something to a small group or individual. Perhaps the teacher will demonstrate to the whole class and you will be reinforcing this by repeating as required on an individual level. Perhaps you will be showing a specific technique or skill, such as cutting or joining to the child you are supporting.

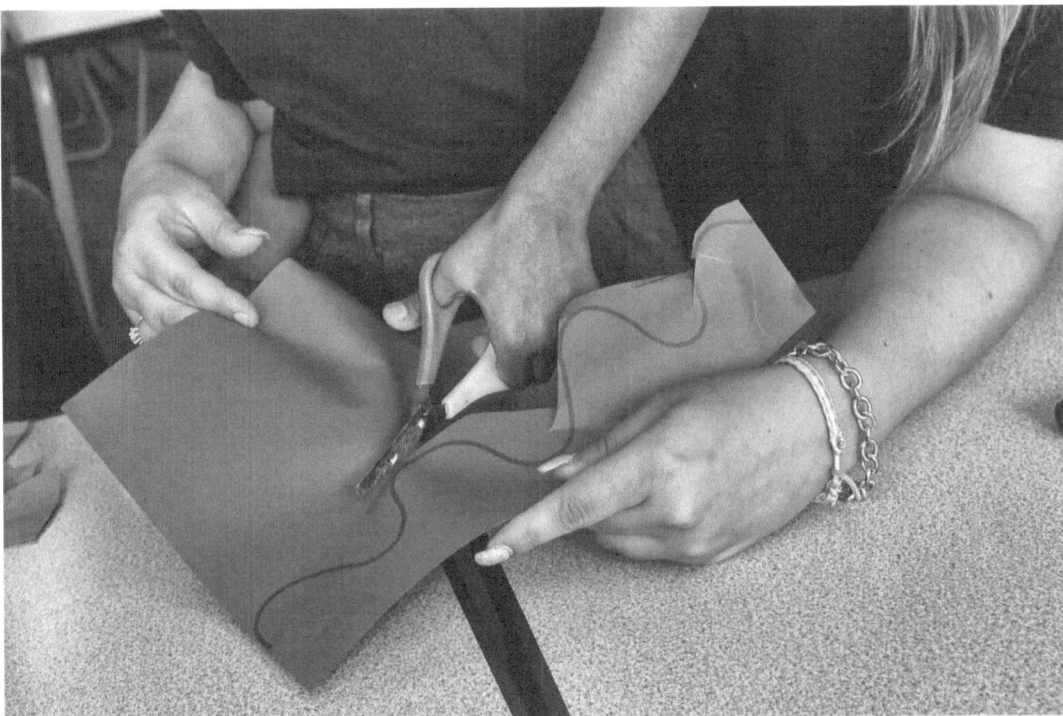

Figure 7.1 Adult-guided cutting

Preparing for demonstrations

- Think about what you are going to do and say in advance.
- Practise the particular technique to be taught until you are confident doing it yourself – teachers often do this.
- Break down the process into smaller stages to aid understanding.
- Think about the possible tricky points for pupils – for example, holding the equipment the right way.
- Choose the most important points to tell pupils. How will you help them remember?
- Plan some questions which will challenge pupils' understanding appropriately.
- Find out if there are any health and safety issues.
- Have the equipment and materials ready and well laid out.

When you are demonstrating:

- Demonstrations are interactive, not one-way dialogues.
- Keep the group size small.
- Position yourself so that everyone can see you, that they can sit or stand comfortably and not get distracted. (Ask yourself if it would be better for them to be behind you, looking over your shoulder rather than seeing a 'mirror image'; e.g. when threading a sewing machine.)
- Explain each step clearly and simply as you demonstrate. You are modelling good technique.

- Work much more slowly than usual; provide a running commentary to point out the detail of each stage.
- Reinforce the salient points. Repeat them and try to explain why things are done in a particular way. Emphasise important points by drawing pupils' attention to them – 'Do you see how this . . .?' – then pause and let the pupils look.
- Make sure the pupils pay attention; involve them by asking questions, make them touch, handle, look, smell and listen in order to to get a better understanding (for example, if mixing dough and you wish to reinforce the point about how wet/dry it will be).
- Keep eye contact with the group repeatedly and check they understand. Involve them if you can, for example by holding things, measuring and choosing materials or tools.
- Review what you have done at the end with the students. If they are going to carry out the task immediately, make sure they all know the first couple of steps.

Talking to pupils while they are designing and making

One of the real benefits of practical activities is the opportunity for support assistants to speak to pupils as they work through their tasks. Much of this talking seems like informal chatting but it can make a significant contribution to their learning, particularly for those pupils who find it hard to speak up in whole-class situations. Talking to pupils, intervening with questions, providing praise and encouragement, and asking their opinions makes a real difference to learning and success.

Questions you can ask pupils

Well-designed questions lead pupils from unsorted knowledge to organised understanding. Questioning is an area that is characterised by a great deal of instinctive practice; such questions allow you to reflect upon, analyse and develop the effectiveness of what you do. Effective questioning is not just a matter of planning which questions to ask, but also planning to stage or sequence those questions so that they guide the pupils toward key lesson objectives. Questioning pupils with SEN should be done in an atmosphere where pupils feel secure enough to take risks: try to allow thinking time, use 'wait time' and value all answers given. Give credit for trying, and maintain positive eye contact and body language to encourage pupils.

The wording of questions becomes acutely important when working with pupils with learning difficulties. Avoid questions such as, 'Can you explain . . .?', and 'Are you able to describe . . .?' Pupils with SEND may interpret them literally and answer yes/no. Replace them with, 'Explain to me', 'Describe to me . . .'. Avoid negative phrases such as, 'I'm not clear about . . .'. Use, 'Tell me what you mean . . .'. Sometimes it may be appropriate to say, 'I think this is what you're saying . . .'.

It is helpful to refer to one pupil's answer to another pupil to generate discussion. Encourage pupils to answer each other's questions: 'What do you think about that?' Encourage them to add to and challenge the answers provided by others. They should listen to each other's questions and answers as well as yours. This develops confidence, social interaction and cooperative working.

Looking at existing products

One essential learning activity in D&T is to investigate and evaluate existing products to inform and explore new design ideas. Looking at existing products is a rich source of talking and questioning. Don't be afraid to express an opinion or share your experiences which might be relevant to what the pupils are doing: 'They have just started recycling bottles and glasses in my area, I think it's a good idea because . . .'; 'I tried a new bread from the supermarket this week. It had apricots and walnuts in it, and that gave it an interesting flavour . . .'.

When thinking about questions for product evaluation:

- Show pupils how to formulate their own questions, for example from a list provided: 'What questions could we ask about this product?'
- Ask pupils how they feel about the product and for other opinions, to build their self-esteem.
- Have groups working on different sets of questions – match the level of challenge to the ability to move different pupils forward.
- Plan higher order questions where appropriate – moving pupils from factual recall and comprehension to application, analysis, synthesis and evaluation.

Questions to ask to get pupils organised

Encourage pupils to think for themselves. Support assistants should promote independence at all times. Ask questions that encourage pupils to think about what they are doing. Probe for explanation and justification by using further questions to extend thinking that requires more than a one-word answer: 'Explain . . .?', 'Why . . .?', 'What makes you think that?', 'What would happen if . . .?', 'Tell me more . . .':

- What do you need to do first?
- What will you need?
- Where are you going to work?
- Who will you work with?

Knowledge	Comprehension	Application	Analysis	Synthesis	Evaluation
Describing the product:	**Translating, interpreting and extrapolating:**	**Applying to new situations:**	**Breaking down into parts/forms:**	**Combining elements into a pattern**	**According to criteria and stating reasons**
What is it made from?	Do I like it?	What is my reaction to this product?	What is the function of this product/parts of the product?	Would I want to own or use it?	What effect will this product have on people's lives and relationships?
Who is it for?	Is it what I need?	Who might the owner be?	What do people think of this product?	What would this reveal about me?	Is this a better product than . . .?
When would it be used?	Is it the right size, shape, pattern, colour, smell, taste?	Why might they want to buy it?	Does everyone think this product is a good invention?	What influenced the appearance and the way it works?	Is this a more important invention than . . .?
Where is it used?	Is it safe?	Does it work well?	Who is it for?	How might the design have been developed?	Is this a more appropriate solution than . . .?
How often is it used?	Does it do the intended job?	What ingredients and processes have been used – why?	What assumptions have been made about the people who might use it?	How would you test this to see . . .?	What is wrong with this product?
Which one . . .?	Is it value for money?	Does it do what it was intended to?	Whose needs or wants were possibly considered during designing and making this product?	Propose an alternative solution to the product or part of product.	Why is this product not as popular as . . .?
What is the best one for doing . . .?	What does it cost in relation to the income of the potential users?	Does it look and taste good?	Why is this product like this?	How else would you . . .?	What could be done better or differently?
How much does it cost?	Demonstrate how the product is used.	How well is it made?	What are the motives of the people who design and make it?	Suggest changes you could make to . . .	How good is this product compared to . . .?
Where is it sold?	Explain why this product was developed.	Is it nicely finished?	What is the relationship between this product and . . .?	Develop a list of important features.	What difficulties do users find with this product?
Who designed and made this?	Explain what is meant by . . . (label or product feature)	Is the cost appropriate?	What makes this product distinct from others of its type?	How is this product different from one from five years ago/another culture?	What difficulties do manufacturers have making this product?
How has it been made?	Give an example of . . .	Is it really needed?	Does this product have an identity or image?	How will this product be different in ten years' time?	What negative impacts does this product have on other people?
Where is it from?	Is this the same as . . .?	How much will this product change people's lives?	How has this been achieved?	What would happen if you were to add . . .?	Why have these particular ingredients been chosen?
What sort/type/category of product is it?	What would happen if this product was used for . . . (another purpose)?	Choose the best statements that apply to this product (statements given).	Does the promotion target a particular age group or group of people?	What would happen if you were to make it?	Can it be part of a sustainable world?
What other products are like this?		How is it promoted and packaged?	What are they trying to say about the product?	Why did the designer make it this way?	Where do the ingredients come from?
		Predict what would happen if ..	How are they persuading you?	What happens to it after use? How long will it last?	Is there a problem with side effects?
			What do people believe about this product – is it true?		What else could have been used?

Figure 7.2 Questions to develop learning and thinking

- Who is doing what in your group?
- How long have you got to do this task?

Questions to help when designing

Sometimes, pupils find designing difficult and will need support and feedback to keep them on task. Don't talk to them too much! Try to balance asking the right questions against talking so much that it distracts the pupils from what they are trying to do, or prevents them from sorting things out for themselves. Help pupils to move forward by intervening with open-ended questions that have more than one answer and require extended responses. Use speculative 'what if' questions. Use 'why' as the opening to questions.

Avoid making decisions for the pupils. Give them opportunities to make choices and decisions for themselves. This encourages independent behavior and thinking amongst them.

Reinforce the use of key technical vocabulary, such as 'ideas', 'model', 'user', 'purpose' and 'design criteria'. Sometimes it will be helpful to simplify the language that is used, and to help pupils by recording their answers and designs for them or with them. They should still retain ownership of the work.

- Who is it for?What do they need?
- What do you want to find out?
- How will they use it?
- What else is like this?
- What did your research tell you?
- What size does it need to be?
- What else could you use?
- What else would work?
- What would happen if you . . .?
- How are you going to make it?
- How could you try it out before you make or model it?
- How could you . . .?
- What do we call it when we . . .?

Questions to help when making

In practical lessons, when the pupil needs to manipulate specialised equipment, the support assistant should work *under the direction of the pupils*. Thus questions will often be, 'What would you like me to do first?'

Where pupils are able to work more independently, listen carefully to their answers: it will tell you whether they understand and are able to get on with the task. Use questions to develop self-help skills. Avoid creating dependency. Intervene to reinforce key points and make learning explicit for pupils: 'Why do we use oven gloves to take out the tray?' (instructions); 'Why do you like the green one?' (opinions); 'I think that you have sewn that very carefully' (feedback); 'Do you think you will need to add some more water?' (advice); 'Don't forget to . . .' (reminder); 'Look at the way this is glued together' (pointing out); 'What do we call this kind of joint?' (reinforcing technical vocabulary).

- What tool would be best for that job?
- How are you going to join those pieces together?
- What could you use?
- How could you make those the right size?
- How are you going to decorate it/finish it?
- How could you make that part look better?

Supporting practical work

A high proportion of time in D&T lessons is spent doing practical work, and support assistants play an integral role in preparing materials, demonstrating how to do something and talking to children while they are working. Watching and listening is very important – it will help to identify what is really going on, how much help individuals need and what kind of help is appropriate.

D&T is about developing pupils' ability to design and make things by themselves, so don't carry out tasks *for* them, unless this has been deemed appropriate. The support assistant's role is to facilitate independent learning and to enable the pupils to do as much as possible for themselves. Pupils who lack self-confidence will sometimes plead with you to do it for them – don't! Guide pupils by talking them through the task, or allow them to instruct you so that you carry out tasks on their behalf. They will need more support if they are doing something for the first time.

Often it will be helpful to segment the task into small steps, to simplify the language used in the work, and keep the pupil on task with feedback and prompting.

Working alongside the teacher in practical lessons

- Ask the teacher if the practical activity needs to be adapted for the pupil you are supporting, for example if there is special equipment you can use.
- Reinforce the classroom routines for setting up work and clearing away.
- Don't do things for pupils that they can do themselves, such as clearing up.
- Labelling cupboards with a digital photo/name label helps pupils to access the equipment they need independently.

- Encourage the pupil to remember where resources are kept.
- Remind pupils how to work safely and hygienically.
- Encourage pupils to work tidily.
- Encourage pupils to work accurately.
- Observe pupils' posture, position and handling of equipment, and correct where needed.
- Ensure the class has equal access to materials and equipment.
- Use the correct names for tools and equipment, and processes to reinforce key vocabulary.

Helping support staff to learn about D&T so that they can support effectively

D&T is a dynamic subject, and teachers are required to constantly update their knowledge and skills to keep abreast of the breadth of the subject and new developments within it. Support assistants similarly will have different strengths and requirements for professional development.

Case study

Katrina, a support assistant, was working in food lessons supporting a pupil with cognitive and learning difficulties. She did not feel confident in her understanding of nutrition and food science, so she used a free (FutureLearn) 6 hour online training course on nutrition and worked through information on a couple of websites to bring her knowledge up to date. She also joined the food teachers for a free webinar provided as part of the Food Teacher Professional Portfolio programme. She then discussed with the teacher how this applied to the lessons for that term.

Case study

Jack, a support assistant, joined a D&T team training day on ICT in D&T, so that he could become familiar with the latest CAD/CAM software that the pupils were using. Jack learned how the software can be used for designing and making. He then discussed with the staff how to support pupils with learning difficulties when they are using ICT.

Summary

Support assistants are a rich resource when used effectively and they can be invaluable to the pressed D&T teacher. They often enjoy supporting design and technology lessons because the practical nature of the subject makes it easy to join in a classroom situation and provide individual support to those who need it, relatively inconspicuously.

Specialist support may be useful for:

- simplifying the language used in the work;
- presenting the project in smaller steps and guiding the pupil through tasks;
- keeping the student on task with support and feedback;
- writing and recording for the student;
- helping the pupil with specialist equipment;
- helping with personal organisation or with physical or medical needs.

Support assistants are used most effectively where:

- Careful consideration has been given to where the help could be used best.
- The teacher and the support staff are able to discuss and plan strategies for working together.
- The pupil is clear about why the extra help is given.
- The support staff are involved in providing feedback on, and assessment of, the pupil's work.

8 Real pupils in real classrooms

This chapter introduces some case studies of pupils and the strategies their D&T teachers use to support them in getting the most from the D&T curriculum. You may like to use these case studies as a reference point or to assist you during CPD sessions with your department.

Special contributions from D&T in your school

- What special contributions do you look for from D&T to balance the curriculum for your pupils with SEND?
- How can you further promote the value of D&T for pupils with SEND to your colleagues and pupils' parents? Review (objectively, as an outsider) how others perceive the nature of the work and achievements by pupils with SEND in D&T.
- Consider the range of SEND you experience across your school and the problems these may present for students' learning in D&T. What solutions and support can you offer?

Case studies

The case studies that follow introduce:

- Kuli in Year 8, who has a hearing impairment.
- Harry in Year 7, who is dyslexic.
- Megan in Year 10, who uses a wheelchair.
- Steven in Year 9, who has emotional, behavioural and social difficulties.
- Matthew in Year 9, who has cognitive and learning difficulties.
- Bhavini in Year 9, who is visually impaired.
- Susan in Year 10 who has complex difficulties, including autistic spectrum disorder.
- Jenny in Year 7 who has Down's Syndrome.

Kuli

Kuli has significant hearing loss. He has some hearing in his right ear, but he is heavily reliant on his hearing aid and visual cues ranging from lip reading to studying body language and facial expression to get the gist and tone of what people are saying. He often misses crucial details. Reading is a useful alternative input, and his mechanical reading skills are good, but he does not always get the full message because of language delay. He has problems with new vocabulary and with asking and responding to questions.

Now in Year 8, he follows the same timetable as the rest of his class for most of the week but he has some individual tutorial sessions with a teacher of the deaf to help with his understanding of the curriculum and to focus on his speech and language development. This is essential, but it does mean that he misses some classes, so he is not always up to speed with subject matter.

He has a good sense of humour, but appreciates visual jokes more than ones that are language based. He is very literal and is puzzled by all sorts of idioms. He was shocked when he heard that someone had been 'painting the town red' as he thought this was an act of vandalism! Even when he knows what he wants to say he does not always have the words or structures to accurately communicate what he knows.

Everyone is very pleasant and quite friendly to him, but he is not really part of any group and quite often misunderstands what other pupils are saying. He has a support assistant which again marks him out as different. He gets quite frustrated because he always has ideas that are too complex for his expressive ability. He can be very sulky and has temper tantrums.

Strategies

- Kuli needs to know what is planned for upcoming lessons so that he can prepare the vocabulary and get some sense of the main concepts so he can understand what is being said.
- A support assistant will find ways of displaying information visually, using drawings, pictures, signs, symbols, sign language, mime, animations on the computer, etc.
- Make sure demonstrations and explanations are seen and understood.
- Try to reduce the general hum of noise in the classroom, particularly during practical lessons.
- Present one source of information at a time: it is hard for him to focus on what is being said at the same time as looking at a book or watching what is being written on a board.
- Phrase questions carefully and always say his name beforehand directing them towards him.

Project example – Alarms

The teacher introduced a task that he felt would engage the interest of all of the pupils: to design and manufacture alarms that could be used in a real-life situation.

The teacher told the pupils that they could choose their own inputs and outputs for the alarm, and asked them to decide for themselves on a purpose for their alarm. The added bonus of this project for Kuli was that the teacher suggested it would be good if he were able to make an alarm for people in his situation, with significant hearing loss. This motivated him a lot.

Kuli had to develop a specification for his product by examining existing alarms and referring to his own proposals. At this point, the teacher was careful to check Kuli's understanding by posing focused questions to him. Kuli decided to make a plant dryness indicator; when the soil around the plant becomes dry, the light-emitting diode (LED) is activated.

The pupils used a systems and control simulator program to help them start the project. Next, the teacher gave the class a generic printed circuit board (PCB) so that they could insert their own inputs and outputs. Kuli found it helpful to use a larger PCB that had wide tracks and pads. In the following lessons, the pupils added components to their PCBs and learned how to use a multi-meter to examine the soldering. The teacher provided differentiated worksheets to support Kuli during the manufacturing processes. He ensured that he gave specific feedback to Kuli as his ideas developed.

As the pupils had used a generic PCB, this meant that they did not have to manufacture individual circuit boards and spent only a small amount of time on fault-finding. This freed up more lesson time to concentrate on designing.

The teacher thought of ways to capture the pupils' interest. For example, when he gave them the task of designing casings, he decided to allow them to choose whichever medium they felt most confident working with. He suggested they could, for example, make models in card, or through using CAD or traditional graphics techniques.

The teacher had undertaken this project so that the pupils could carry out the design and make activity within a defined framework whilst also having the freedom to design something they would find useful. He noticed that this increased motivation in Kuli and others in the class. It allowed the pupils to demonstrate their own ideas both to the teacher and to each other. This gave Kuli the confidence to work on his own initiative and helped secure his motivation and concentration.

Harry

Harry is a very anxious little boy and, although he has now started at secondary school, he still seems to be a 'little boy'. His parents have been very concerned about his slow progress in reading and writing and arranged for a dyslexia assessment when he was eight years old. They also employ a private tutor who comes to the house for two hours per week and they spend time each evening, and at weekends, hearing him read and working on phonics with him.

Harry expresses himself well orally using words which are very sophisticated and adult. His reading is improving (RA 8.4) but his handwriting and spelling are so poor that it is sometimes difficult to work out what he has written. He doesn't just confuse *b* and *d* but also *h* and *y*, *p* and *b*. Increasingly, he uses a small bank of words that he knows he can spell.

His parents want him to be withdrawn from French class on the grounds that he has enough problems with English. The French teacher reports that Harry is doing well with his comprehension and spoken French, and is one of the more able children in the class.

Some staff get exasperated with Harry as he is quite clumsy, seems to be in a dream half the time and cannot remember a simple sequence of instructions. He has difficulty telling left from right and therefore is often talking about the wrong diagram in a book or out of step in PE and sport. 'He's just not trying', said one teacher, while others think he needs 'to grow up a bit.'

He is popular with the girls in his class and recently has made friends with some of the boys in the choir. Music is Harry's great passion, but his parents are not willing for him to learn an instrument at the moment.

Strategies

- Staff should talk to the parents about Harry's lack of confidence.
- Provide Harry with strategies for distinguishing left from right.
- Find out how he has learned things in the past and see if similar strategies would work in the classroom.
- Investigate the possibility of Harry using a computer with a spellcheck function at home and school to cope with orthographic and spelling difficulties.
- Offer Harry lots of praise at appropriate times.
- Allow Harry more time to complete practical work if needed.
- Focus on Harry's strengths – talk about and discuss designs, products, making and evaluating rather than writing about them.
- Make use of pictures, plans, flowcharts and visual instruction sheets with digital images of processes.

- Provide templates for Harry to complete and give him key words to use.
- Practise names of equipment and processes with Harry to reinforce key vocabulary.
- Show Harry how to annotate neatly.

Project example – Design and make a snack bar

The teacher set Harry a design and make assignment that required him to think about how a food product is created, see how a recipe can be altered to suit different tastes and to learn the importance of nutritional content.

Luckily, Harry enjoys the practical aspects of design and technology: the reading and writing have a real purpose in the lesson as they lead to making something edible! Harry enjoys the success of seeing something he has made and this has given him the confidence to record his ideas and evaluations in a simple way.

Through group discussion and teacher-led prompts, Harry was encouraged to devise three criteria against which to evaluate his product:

- it should taste good;
- it should 'do you good';
- it should be well wrapped, to travel in a lunch box.

The teacher provided a range of commercially-produced snack bars for the groups to unwrap, examine, discuss, taste and evaluate nutritionally and against the set criteria for the assignment. It was more important to discuss and describe the products than record thoughts at this point, and the teacher frequently praised Harry for his contributions. Harry was given a chart template to fill in the information collected and a simple word bank to use in doing it.

The teacher talked about the nutritional value of foods – classifying the food groups, the balanced plate and the healthy-eating model. He was assisted in putting examples of foods into the right section of the plate by a support assistant.

During the focused practical tasks, Harry was shown how to make both biscuit- and cake-based products and how to add a given variety of ingredients to alter the taste and nutritional content of the product. Harry practised naming and weighing ingredients, using a variety of equipment, including electronic food mixers. He tried different 'toppings' and decoration on the snack bars. The teacher asked him to use matching cards to show the sequence of actions after he had made the recipe.

The teacher devised a chart to simplify Harry's choice of ingredients and he cooked his design. He had to focus on the steps in the recipe, with teacher support, to make sure he was doing one step at a time. Some of the product was frozen for further evaluation.

Throughout the activity, the teacher enabled Harry to participate at his own level through careful positioning of support staff, simple visual recipes and actual ingredients, to enable Harry to indicate his choice and 'use' worksheets and recipes to assist working as independently as possible.

Megan

Nicknamed 'Little Miss Angry', everyone knows when Megan is around! She is very outgoing, loud and tough. No one feels sorry for her – they wouldn't dare! Megan has spina bifida and needs a wheelchair and personal care as well as educational support. She has upset a number of the less experienced classroom assistants who find her to be very difficult to work with at times. Some of the teachers like her because she is very sparky. If she likes a subject, she works hard – or at least she did until this year.

Megan has to be up very early in the morning for her parents to help get her ready for school before the bus comes at 7.50 a.m. She lives out of town and is one of the first to be collected and one of the last to be dropped off at the end of the day. She threfore has a longer school day than many of her classmates. Tiredness can be a problem for her, as everything takes her so long to do and involves significant effort.

Now she is 15, she has started working towards her GCSEs and has the potential to get several A-to-Cs, particularly in Maths and sciences. She is intelligent but is in danger of becoming disaffected because everything is so much harder for her than for other children. Recently she has lost her temper with a teacher, made cruel remarks to a very sensitive child and turned her wheelchair round so she sat with her back to a supply teacher. She has done no homework for the last few weeks saying that she doesn't see the point as 'no one takes a crip seriously'.

Strategies

- The school needs to identify those staff that Megan is on good terms with and make sure she spends time with the people she respects.
- Urgent support is needed to minimise the physical effort involved for her in writing and recording.
- Staff need to discuss things with her instead of talking behind her back and give her some respect.

- The school needs to establish ground rules with Megan about acceptable behaviours.
- Megan, her parents and school staff should consider and discuss reducing the number of subjects she is taking.
- Staff should have higher expectations of her and demonstrate this.
- Rotate support assistants to reduce the stress placed on them while working with Megan.
- Talk to parents regarding health – has there been some deterioration? Problems at home?
- Organise practical work carefully and thoughtfully, taking account of Megan's full range of additional needs.
- As much independence as possible should be allowed to Megan.

Project example – Corporate identity

Souvenirs and collectables, e.g. T-shirts, 3D signs and models are used to promote events, pop stars, cartoon characters and even schools. Design and make a coordinated range of promotional products for a special occasion or a corporate client.

A teacher re-negotiated this project with Megan to ask her if she would like to work with a group that was going to be more challenged. Megan enjoys D&T, though she does become tired during practical sessions. The room is adapted for a wheelchair user and she can access nearly all of the equipment independently. She enjoys food technology because it helps her to work on skills for looking after herself in the future and she is fiercely independent.

The group were going to work directly with an outside client, by e-mail and video conferencing to develop a set of promotional products for an event that was coming up for them.

The teachers supported the group and ensured that they had help when they required it, but once the task was outlined they were left to manage the project for themselves, coming to the teacher with requests when they needed to know something, and reporting back at regular intervals. The group decided on team roles and negotiated their work directly with the company, working independently as far as possible. Megan and the group had to work with a specification requiring that the products were innovative, but also thoroughly tested and of marketable quality. They also designed for an event that was not familiar to them and required rigorous research on their part.

They were ambitious in the range of products they chose and the ideas they presented. As a result, they took greater risks and coped with a greater number of variables. The use of CAD/CAM meant Megan produced some high-quality products from a design that the pupils worked on as a group.

One of the major contributions that Megan made, as well as helping to design and make the products with the group, was to record the events with a digital camera and put together a powerpoint and video presentation to the company, showing their prototypes and research.

Megan enjoyed the challenge and, as a result of working with an outside company, she felt accepted and respected for her ideas and skills.

Steven

'Stevie' is a real charmer – sometimes! He is totally inconsistent: one day, he is full of enthusiasm; the next day, he is very tricky and he needs to be kept on target. He thrives on attention. In primary school, he spent a lot of time sitting by the teacher's desk and seemed to enjoy feeling special. If he sat there he would get on with his work but then, as soon as he moved to sit with his friends, he wanted to make sure he was the centre of attention.

Now in Year 8, Steven sometimes seems lazy – looking for the easy way out, but at other times he is quite dynamic and has lots of bright ideas. He struggles to work independently and has a very short attention span. No one has very high expectations of him and he is not about to prove them wrong.

Some of the children don't like him because he can be a bully, but really he is not nasty. He is a permanent lieutenant for some of the tougher boys and does things to win their approval.

He is a thief but mostly he takes silly things, designed to annoy rather than for any monetary value. He was found with someone's library ticket and stole one shoe from the changing rooms during PE.

Since his mother began a relationship with a new partner, there has been deterioration in his behaviour; Steven has also been cautioned by police after stealing from a local DIY store. He has just been suspended for throwing a chair at a teacher, but staff suspect this was because he was dared to do it. He certainly knows how to get attention.

Strategies

- Steven would benefit from a structured programme with lots of rewards – certificates, merits, etc.
- Praise Steven so that he overcomes his negative self-image.
- Use success to build confidence, for example, use CAD and electronic portfolios to produce a 'professional looking' result in tasks.
- Outline information in short, easily digestible, chunks.

- Change activity frequently, in order to keep Steven engaged.
- Allow Steven some individual responsibility while completing tasks.
- Give Steven opportunities to express his opinion about products (that he is interested in), help him to feel empowered that he is able to design something to make a difference.
- Give positive support to promote independence.
- Give regular feedback and opportunities for Steven to his improve work.
- Provide a supportive work group where relevant.

Project example – Handheld technologies

Fortunately, Steven really enjoys D&T because it gives him an opportunity to experience success. The teacher began this project by showing Steven examples of handheld technology, encouraging him to be critical about the products and to think of ways that they could be improved. Leading questions were asked to encourage him to reflect critically.

Pupils were given the opportunity to use a computer rather than paper to produce a portfolio, and Steven chose this option. Using an electronic portfolio helped to keep him on task, although many other strategies are also required alongside this.

Steven used the internet to find and download examples of handheld technology and was able to give his opinions on the items in a very basic way. For this project he used examples of existing products to inform his design ideas. The designs had to be presented in isometric projection and Steven used CAD to do this. He put forward some design ideas produced with CAD and evaluated these in a simple manner. At this point he needed visual support (instruction sheets) to help him with CAD software as he does not cope well with failure and it was essential to keep him on task. A pictorial guide had been produced using screen grabs that Steven could follow in order to produce his design. Using the pictures, instead of lots of writing, really helped him to work independently. The teacher also had a laptop linked to a projector and it was possible to illustrate processes step by step so that the pupils could follow the teacher's instructions when they got stuck.

Steven went on to develop his work and propose a final design. This part of his idea development was basic, but showed that he has some understanding of this aspect of the design process.

Next, the class were shown how to make a product model using Styrofoam. This included the use of a CAM machine to make the buttons. Success at this stage was ensured by managing the classroom, which involved assigning different areas of the room to different activities. Pupils were given regular inputs to assist practical work and support was on hand whenever they came to use machinery. During the making stage, Steven needed to be closely monitored to ensure he was doing the work required of him and not disturbing anyone else. This was achieved through constant praise and reassurance. It was also necessary to group him with pupils who would support him, rather than antagonise him.

Steven completed his product and kept a process diary or step-by-step guide. As products were being constructed, Steven was encouraged to keep a diary so that he understood which machines were used for various processes. This is a forerunner to pupils planning for themselves. There was a simple evaluation at the end of the project.

In order to maintain interest in the project, each aspect of the work was treated as an end in itself and pupils were given feedback in the form of grades and verbal feedback. Steven was given clear guidance about how his work could be improved and he re-submitted the work at the end of the project for summative assessment.

This was a major achievement for Steven as it was one of the few pieces of work that he managed to complete that year. He said he was really proud of himself, which was good to hear. The key to success was positive support and regular feedback.

Matthew

Matt is a very passive boy. He has no curiosity, no strong likes or dislikes. One teacher said, 'He's the sort of boy who says yes to everything to avoid further discussion, but I sometimes wonder if he understands anything.'

Now in Year 9, he is quite an introvert. He knows all the children, but does not feel uncomfortable with them and is always on the margins. Often in class he sits and does nothing, he just stares into space. He is no trouble and indeed if there is any kind of conflict, he absents himself or ignores it. No one knows very much about him as he never volunteers any information. In French class, he once said that he had a dog, and one teacher has seen him on the local common with a terrier but no one is sure if it is his.

He does every piece of work as quickly as possible to get it over with. His work is messy and there is no substance to anything he does, which makes it hard for teachers to suggest a way forward, or indeed to find anything to praise. Matthew often looks a bit grubby and is usually untidy. He can be quite clumsy and loses things regularly, but does not look for them. He does less than the minimum.

He is in a low set for maths but stays in the middle. He has problems with most humanities subjects because he lacks empathy and no real sense of what is required. When the class went to visit a museum for their work on war, he was completely unmoved. To him, it was just another building and he could not really link it with the work the class had done in history.

Strategies

- Get parents/carers involved to find out if he has any enthusiasms at home.
- Involve him in pair work with a livelier pupil who will 'gee him up' a bit.
- Set up situations where he can make a contribution.
- Set up some one-to-one sessions with a support assistant where he is pushed to respond.
- Get him using technology to improve the appearance of his work, perhaps in a homework club after school.
- Segment D&T projects into small tasks to give regular rewards of success.
- Ask Matthew to adapt a design, rather than starting from scratch with a new idea.
- Give him support during design and evaluation stages of projects so that he does not lag behind other pupils.

Project example – Design and make a light or lantern

The basic circuit for the light was made and used in science lessons.

The first part of the D&T project was a teacher-led focused practical task to show how to make a timber framework for the light/lantern. The dimensions of the frameworks varied to suit the circuits, which had been mounted onto reclaimed materials. The task was broken down into smaller steps. Matthew received as much support as he needed to do this, so that the product would not be 'lost' because he could not manipulate the timber and card, or measure accurately. Matthew was shown the end result before he started so that he knew what he was aiming to achieve.

The design and make brief was presented to the group when they had made a framework and installed a circuit. Product evaluation activities were used to examine existing torches/lights and lanterns and discuss their purposes. The group identified the fact that they were all comfortable to carry around, and that they could be switched on and off easily. They also noted that the circuits were protected by the case, and that the bulb was protected by a transparent lens.

Matthew was asked to create and annotate the drawing, so that he could make a handle for the light. If he had not been able to draw a design, he could have indicated a preference, but he might not have been able to think of how to make his design work.

He was also given attractive translucent paper to clad the framework with and protect the circuit and bulb. He was asked to check that he could still reach the switch, but he did not have to produce a design to show how he would achieve this.

Matthew tested the products for ease of transport and use of the switch. He evaluated the lantern, first with no support, then as part of a teacher-led writing activity.

Making sure that Matthew had a good end product helped him to feel positive about D&T. Provided some form of informal recording takes place, a lot of adult help does not inflate the assessment of the child's level of achievement. In order to let children working at Matthew's level take part in these interesting class activities, a lot of adult intervention may be needed, allowing the child responsibility for the design and how it is made.

Bhavini

Bhavini has very little useful sight. She uses a stick to get around school and some of the other children make cruel comments about this, which she finds very hurtful. She also wears glasses with thick lenses which she hates. On more than one occasion, she has been knocked over in the corridor, but she insists that these incidents were accidents and that she is not being bullied. However, her sight is so poor that she may not recognise pupils who pick on her.

Bhavini uses a certain amount of specialist equipment, such as talking scales in food technology but, now in Year 9, she is always conscious of being different. Her classmates accept her but she is very isolated as she does not make eye contact, or see well enough, to find people she knows to sit with at break. She spends a lot of time hanging around the support area. Her form tutor has tried to get other pupils to mentor her, or to escort her to humanities, which is in another building, but this has bred resentment. She has friends outside school at the local 'Phab' club (Physically handicapped/able-bodied) and has taken part in regional VI Athletics tournaments, although she opts out of sport at school if she can. Some of the teachers are concerned about health and safety issues and there has been talk about her not applying herself in science.

She has a reading age approximately three years behind her chronological age and spells phonetically. Many of the teaching strategies used to make learning more interesting tend to disadvantage her. The lively layout of her textbooks, with cartoons and speech bubbles, is a challenge to her to read. Even if she has a photocopy of the text enlarged, she cannot track which bit goes where. At the end of one term, she turned up at the support base asking for some work to do because, 'They're all watching videos'.

Strategies

- Her isolation is the key factor in her well-being and needs addressing most urgently. Make sure she can work on the same projects as the rest of the class.
- She needs to be put in groups with different pupils who will not overwhelm her.
- Work on spelling – core vocabulary and spelling patterns which are not phonic.
- Utilise specialised equipment such as talking weighing scales or a talking microwave to encourage independence.
- Produce individual resource material for her that is uncluttered and well spaced.
- Use technology to record instructions for processes and recipes.

Project example – Adapt a bread recipe to develop a product suitable for a teenager

D&T presents particular problems for pupils with visual impairment, as they need a bank of practical skills before being able to design effectively. For Bhavini, it is difficult to think beyond the fact that food products just magically appear on the shelf. It is quite a surprise for her to find out about the complex procedures that are in place to develop a food product. The focus of teaching is on developing much needed life skills, such as finding your way, getting organised independently, getting equipment out and working safely. Thus, D&T plays a rich part in the curriculum for Bhavini and she enjoys learning new things all the time.

The teacher sets up a yeast experiment with balloons on test tubes, to show the right conditions for yeast to grow. Through discussion, pupils identify that there are certain conditions needed to help yeast grow in order for carbon dioxide to be released. Bhavini is able to feel the size of the balloons to identify which test tube is working well. Simple diagrams could be drawn on German film.

The teacher plans a focused practical task for making simple bread recipes. This involves listening to instructions, recapping the yeast experiment and familiarisation of the work area and the talking equipment. The shaping and forming of dough helps pupils to develop a variety of design ideas.

These focused practical tasks are 'mini making' activities designed to develop skills, add to knowledge and understanding in a practical way, extend the pupils' experience of different types of breads and create an enjoyable activity that has an end product.

Most pupils will not be able to design and make a new bread product without trying out some existing recipes first. It is quite difficult for Bhavini to complete simple tasks like rubbing in, or adding the correct amount of water so that the mixture is ready to squeeze together into a ball before you knead it. Kneading is fun for most children, but she lacks the ability to apply pressure and this has to be practised.

The teacher also sets up a tasting session of existing breads for pupils. A variety of different breads are used to develop the organoleptic qualities concerned with tasting, but also to enhance pupils' knowledge of the types of ingredients that can be used to make interesting bread. A strict procedure transpires which involves recording the sample name, smell, appearance, taste and texture, followed by a drink to cleanse the palate. Bhavini records her results on prepared sheets in Braille or print. This information is used to help her design development.

When designing, Bhavini is encouraged to record her design ideas verbally or to use Wikki Stix®, German film or paper. Less time is spent on recording designs at this stage. She is encouraged to give as much information as possible. She must try to write a simple specification and a list of ingredients needed. She should be supported in doing this if she needs it. During Years 7 and 8, the details for products are simple in order to make sure she is able to achieve the requisite specification in her practical work. This also reinforces her design work – she can feel the real product and compare it to the raised diagram (using Wikki Stix®) – not an easy concept.

Bhavini is able to draw in reasonable detail on paper. As you can see, the work is fine for a visually-impaired pupil. Improvements could be made, but only by using the computer to enhance the quality of the design and written work. However, Bhavini's practical skills are good and she is able to take into consideration the working characteristics of the ingredients used and to shape, form and finish the bread appropriately. When her ICT skills improve she will improve her work, and probably be able to do entry level/GCSE in due course.

Like most pupils, Bhavini finds evaluating very hard. Depending on how many weeks are available she may have several attempts at developing the design and recording any changes made as she progresses. Evaluation sheets are provided with structured questions and supporting information.

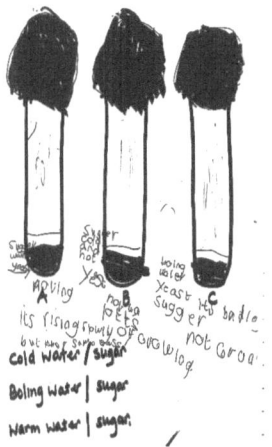

Figure 8.1 Drawing of test tubes

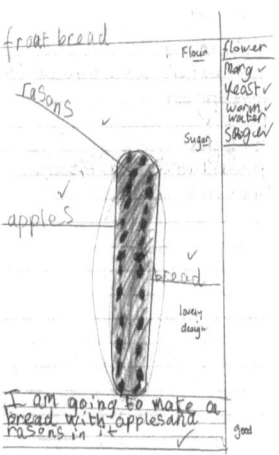

Figure 8.2 Bread

Figure 8.3 Wikki Stix®

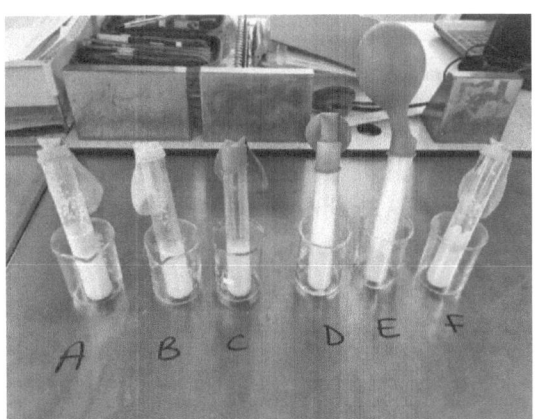

Figure 8.4 Yeast experiment

Susan

Susan is a tall, attractive girl who has been variously labelled as having Asperger's and 'cocktail party' syndrome. She speaks fluently but usually about something totally irrelevant to the task at hand. She is very charming and her language is sometimes quite sophisticated, but her ability to use language for school work in Year 10 operates at a much lower level. Her reading is excellent on some levels, but she cannot draw inferences from the printed word. If you ask her questions about what she has read, she looks blank, echoes what you have said, looks puzzled or changes the subject – something she is very good at.

She finds relationships quite difficult. She is very popular, especially with the boys in her class. They think she is a laugh. There have been one or two problems with some of the boys in school. Her habit of standing too close to people and her over-familiarity have led to misunderstandings which have upset her. Her best friend Laura is very protective of her and tries to mother her, to the extent of doing some of her work for her so she won't get into trouble.

Her work is limited. In art, all of her pictures look the same, very small cramped drawings and she does not like to use paint because 'it's messy'. She finds it very hard to relate to the wider world and sees everything in terms of her own experience. The class has been studying *Macbeth* and she has not moved beyond saying, 'I don't believe in witches and ghosts'.

Some teachers think she is being wilfully stupid or not paying attention. She seems to be attention seeking, as she is very poor at turn taking and shouts out in class if she thinks of something to say or wants to know how to spell a word. When she was younger, she used to retreat under the desk when she was upset and had to be coaxed out. She is still easily offended and cannot bear being teased. She has an answer for everything and, while it may not be sensible or reasonable, there is an underlying logic present.

Strategies

- To help her to become more independent, allocate carefully structured individual tasks and achievements – give Susan specific group tasks.
- Encourage her to count to 20 before expressing her opinions.
- Move her away from Laura.
- Provide writing frames and model answers for her to base her work upon.
- Discuss social issues, body language, appropriate behaviour, etc.
- Keep verbal instructions brief and simple.
- Prepare her for changes (such as moving her away from Laura) in advance.
- Make good use of computers and tablets for designing, such as to add logos and features to work, creating evaluation sheets, digital photos and record sheets.

Project example – Design and make a wheeled vehicle

The teacher structured initial tasks to identify the common features and parts of wheeled vehicles and introduce appropriate vocabulary. Susan was asked to collect pictures showing different parts of vehicles and label them for a group display.

The teacher structured focused practical tasks, including making card vehicles from a net, drawing on vehicle parts and adding wheels and axles. The models were then tested by releasing them down a ramp. The performances of the vehicles were compared and recorded and the topic of friction discussed. The class used a graph to project results. Susan made a card vehicle from a standard net. She then tested her vehicle for smooth running down a ramp and recorded her results using RM Starting Graph software. This 'mini-making' focused practical task was useful as pupils were able to 'play' with what they made.

Through a teacher-directed focused practical task, Susan made a frame for the chassis from wood strip, adding axle supports, wheels and axles. A full-size plan of the chassis was given to Susan so that she could check the dimensions of the design for the vehicle by placing it on the plan.

The pupils then went on to design their vehicles. Susan drew a full-size plan of the vehicle to be made for her design and make assignment. She used cm-squared paper and drew the body of the vehicle within the blue frame already drawn on the paper.

Figure 8.5 Vehicle bodywork

Figure 8.6 Vehicle chassis

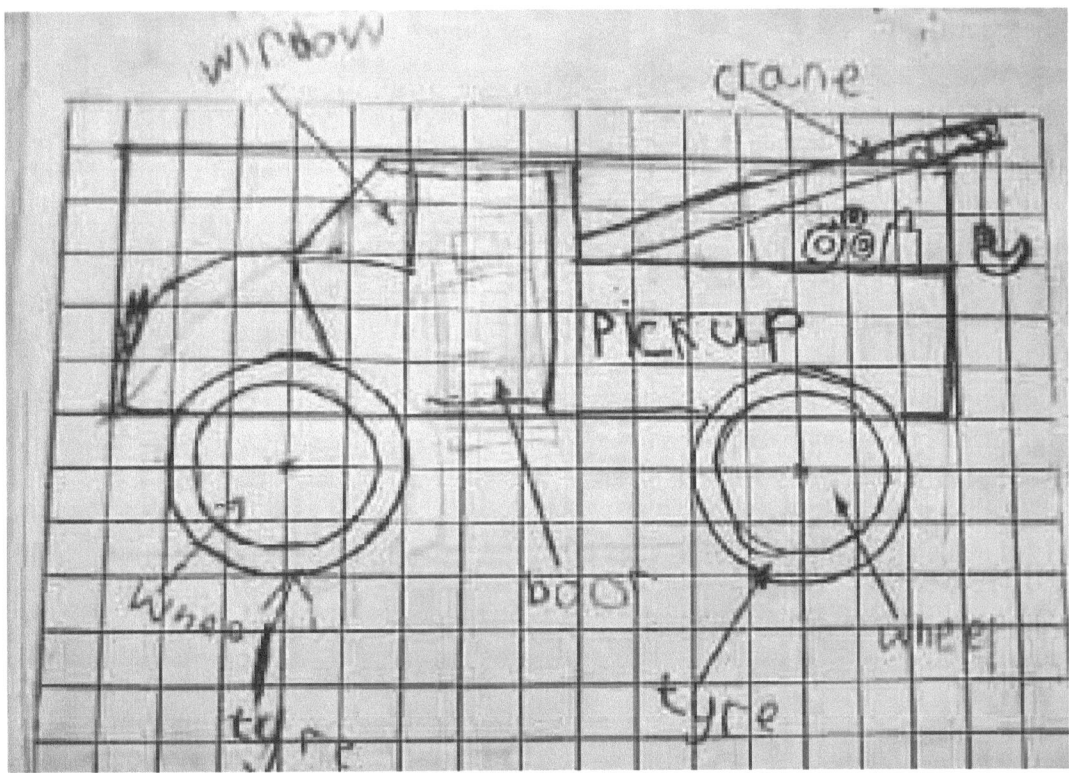

Figure 8.7

All pupils were able to carry out the making tasks successfully, following practice, and worked fairly accurately. Susan, using the technique she had learned earlier in the focused practical task, made a frame for the chassis from a wooden strip. The vehicle body was made from plastic foam and laminated to give the correct depth. Susan made a template from the plan and drew around it onto the foam. She then cut out the shape with a coping saw and sanded the model by hand. The body was painted with acrylic paint and details were added with pens, metallic paper and card. Susan experimented with using ICT to create logos and business names to add to the vehicle.

Finally, the vehicles were tested. Simple evaluations were carried out on the computer using a template and prompt questions and phrases. Susan's final evaluations were basic and required a heavily-structured template with limited choices of phrases from a bank.

Susan required a great deal of structure in the 'vehicles' project. She was able to make simple decisions and choices but found it, in general, very difficult to generate ideas of her own. The introduction of a key, subject-specific, vocabulary encouraged pupils to use the correct terms when speaking and writing.

Task: To make a wheeled vehicle	
I used	*Things that were difficult*
Tools: coping saw, safety rule, lynx jointer, handsaw. Resources: wood strip, card triangles, wooden wheels, dowel, plastic tubing, shiny paper, paper clip, thread.	• Putting the wheels on. • Cutting the body.
What I had to do	*Final evaluation*
1. Draw my design on a piece of squared paper. 2. I made a template of my pick-up. 3. I drew round the template on the foam. 4. I cut the body out with a coping saw. 5. I made the chassis from wood strip. 6. I painted the body of my pick-up and put some details on it. 7. I made the crane out of wood strip and a paper clip. 8. I used Word to make labels. 9. I stuck the body on the chassis. 10. I put the axle supports and wheels on my pick-up.	I am really pleased with my pick-up. It works well and it goes in a straight line. It was fantastic making the vehicle.

Figure 8.8 Evaluation sheet

Jenny

Jenny is in Year 7 and has Down's Syndrome. She is a very confident child who has been cherished and encouraged by her mother, brothers and sisters. She is very assertive and is more than capable of dealing with spiteful comments: 'I don't like it when you call me names. You're cruel and I hate you', but this assertiveness can lead to obstinacy. She is prone to telling teachers that they are wrong!

Jenny has average skills in reading and writing but her work tends to be unimaginative and pedestrian. She enjoys biology, but finds the rest of the science curriculum hard going. She recently started to put on weight and tries to avoid PE. She has persuaded her mother to provide a note saying that she tires easily, but staff know that she is a bundle of energy and is an active member of an amateur theatre group which performs musicals. She has a good singing voice and enjoys dancing.

She went to a local nursery and primary school and fitted in well. She always had someone to sit next to and was invited to all the best birthday parties. Teachers and other parents frequently praised her and she felt special.

Now in secondary school, everything has changed. Some of her friends from primary school have made new friendships and don't want to spend so much time with her. She is very hurt by this and feels excluded. She is also struck by how glamorous some of the older girls look and this has made her more self-conscious.

Strategies

- Talk to parents about diet and exercise and find a way of making Susan feel more attractive.
- Encourage new social groupings in class, so she gets to meet and socialise with other children.
- Pair her up with a child who has better imaginative/empathy skills but weaker literacy, so they can support each other.
- Be aware that too much one-to-one support can be counter-productive: support independence as much as possible.
- Reinforce discussions with pictures and real products; keep listening activities short.
- Use short, clear instructions.

Project example – Design and make a pencil case

The teacher asked the class to work on a project to design and make a textile container or carrier, such as a pencil case, that could be manufactured in batches or as a single item. Pupils developed a basic design that could be varied or personalised for particular clients. The assignment required them to justify their decisions about materials and methods of making. The part that appealed to Jenny was being able to make something that she could take home and use.

The teacher organised a class discussion on the subject, 'Where do new design ideas come from?'. The class used information from designers and older pupils to help develop their ideas. The teacher described to the pupils how designers record their thoughts, design ideas and explorations, for example how they use sketchbooks, moodboards, collages, drawings or collections of inspiring photographs. She showed the pupils good and poor examples of recording and explained that they would need to choose the best methods for their design work.

Jenny was asked to collect pictures, photos and samples of fabric that interested her to put in her sketchbook for designing later. The teacher showed Jenny how she could use these ideas imaginatively: 'Look at this postcard of a beach: you could try to match the colours of the sea'; 'Look at the pattern on this fabric closely: What do you see? What shapes are there?'

During focused practical tasks, the teacher showed the pupils examples of how manufacturing aids (for example, printing blocks and patterns) can be made or used to help with single or batch production. The teacher discussed how to design and make identical parts in a batch. The teacher also made samples of seams, and showed the group possible fastenings. Jenny was given time to practise these to reinforce basic sewing skills, and to make sure the end product was successful.

Jenny generated some ideas during a whole-class discussion and identified possible information sources, such as magazines, from which she took examples of existing products. Her designs did not really move far from copying images that existed already, but she did talk about price for materials and who might buy her kind of pencil case.

The teacher showed Jenny various printing methods she could use to apply colour to her design and she tried some out. She took users' views into account during the testing of materials with a group of other pupils, in order to identify the most suitable fabric to use.

Jenny communicated alternative ideas for pencil cases through annotated sketches; she also showed the design in different colours. She then selected the best combination to use in her final design drawing. The teacher encouraged Jenny to present her ideas during a lesson plenary to the rest of the class, with several other pupils, so that she had a chance to speak to others.

Jenny followed a basic step-by-step plan outlining the order of her work and the equipment to be used. She worked with a variety of materials and components with help from adults. She could choose tools for the job from a limited range.

She was not able to focus her final evaluation well, but she identified what she believed worked well and what could be improved.

Appendix 1.1 SEND legislation

The Children and Families Act: A different landscape

The Children and Families Act 2014 introduced radical changes to the requirements placed on both schools and teachers regarding the education and inclusion of pupils with special educational needs and disabilities. The first major revision of the SEND framework for thirty years, it introduced a new system to help children with special educational needs and disabilities and shaped how education, health and social care professionals should work in partnership with children, young people and their families.

The reforms introduced a system to support children and young people from birth up to the age of 25, designed to ensure smooth transitions across all services as they move from school into further education, training and employment. The reforms give particular emphasis to preparing children and young people for adulthood from the earliest years. This means enabling children to be involved at as young an age as possible in all decisions relating to their learning, therapy, medical treatment and support from social care. The result of this preparation should be that when young people reach the age of 16, they are able to be full and active participants in all important decisions about their life.

There is now an important distinction made between a child and a young person. The Act gives significant new rights directly to young people when they are over compulsory school age but under the age of 25. Under the Act, a child becomes a young person after the last day of summer term during the academic year in which he or she turns 16. This is subject to a young person 'having capacity' to take a decision under the Mental Capacity Act 2005.

Throughout this book the term 'pupils with special educational needs and disabilities (SEND)' is used. A pupil has special educational needs if he or she:

- has a significantly greater difficulty in learning than the majority of others of the same age; or
- has a disability which prevents or hinders him or her from making use of facilities of a kind generally provided for others of the same age in mainstream schools or mainstream Post-16 institutions.

(SEND Code of Practice: DfE 2015)

Section 19 Principles

Central to Part 3 of the Children and Families Act 2014 is Section 19. This section emphasises the role to be played by parents/carers and young people themselves, in all decision making about their SEND provision.

Part C of Section 19 issues a new challenge to schools in that there is a clear expectation that parents and pupils will not only be invited to participate but that they should be supported to do so. This will certainly involve the provision of relevant information to parents but schools could also consider providing other forms of support, both practical support, such as helping with translation, or transport to attend important meetings, and emotional, such as advocacy or pre-meetings to prepare parents and pupils to take a full part in all decisions. Many parents will need only a minimal level of additional support, but others – especially those often portrayed as 'hard to reach' – may require considerably more.

Key questions:

- Do you know how your pupils with SEND and their parents feel about education, and what their wishes are? If not, how can you find out?
- What could you and others in your subject/departmental team do to integrate this information into your planning for and delivery of teaching and learning?
- What more could you do to reach out to parents who may be anxious about or unwilling to engage with school?

The SEND Code of Practice

As the quotation at the start of chapter 1 makes clear, SEND provision is provision that is additional to or different from the high quality, differentiated teaching to which all pupils are entitled. A school's first response to a pupil falling behind his or her peers should be to evaluate the quality of teaching and learning the pupil currently receives in all subjects. The pupil should be identified as having SEND only when the school is confident that all teaching is differentiated appropriately to meet that individual pupil's needs.

Once a pupil is identified as having SEND, schools are required to do whatever they can to remove any barriers to learning and to put in place effective provision, or 'SEND Support'. This support must enable pupils with SEND to achieve the best possible outcomes.

Most schools and academies welcome pupils with a range of vulnerabilities, including special educational needs and disabilities, but may hesitate about including those with significant or complex needs. The reasons behind this reluctance are often a lack of expertise in an area of need, worries about behaviour, and most commonly expressed, concerns about the impact of that pupil's needs on the education of others.

The SEND Code of Practice is very clear that where the parent of a pupil with an education, health, and care plan (EHC plan) makes a request for a particular school, the local authority **must** comply with that preference and name the school in the plan unless:

- it would be unsuitable for the age, ability, aptitude or SEN of the child or young person, or
- the attendance of the child or young person there would be incompatible with the efficient education of others, or the efficient use of resources.

(DfE 2015: 9.79, p.172)

Legally, schools cannot refuse to admit a pupil who does not have an EHC plan because they do not feel able to cater for his or her needs, or because the pupil does not have an EHC plan.

Outcomes

Outcomes are written from the perspective of the pupil and should identify what the provision is intended to achieve. For example, do you think the following is an outcome for a pupil in Year 7 with literacy difficulties?

For the next 10 weeks Jake will work on an online literacy program for 20 minutes three times each week.

It may be specific and measureable; it is achievable and realistic; and it is time targeted, so it is 'S.M.A.R.T' but it isn't an 'outcome'. What is described here is provision, i.e. the intervention that the school will use to help Jake to make accelerated progress.

Outcomes are intended to look forward to the end of the next stage or phase of education, usually two or three years hence. Teachers will, of course, set short-term targets covering between six and 12 weeks and EHC Plans will also include interim objectives to be discussed at annual reviews. So, what would be an outcome for Jake?

> By the end of Year 9, Jake will be able to read and understand the textbooks for his chosen GCSE courses.

The online literacy course would then form a part of the package of provision to enable Jake to achieve this outcome.

The graduated approach

The 2015 SEND Code of Practice describes SEND Support as a cyclical process of assess, plan, do and review that is known as the 'graduated approach'. This cycle is already commonly used in schools, and for pupils with SEND it is intended to be much more than a token, in-house process. Rather it should be a powerful mechanism for reflection and evaluation of the impact of SEND provision. Through the four-part cycle, decisions and actions are revisited, refined and revised. This then leads to a deeper understanding of an individual pupil's needs whilst also offering an insight into the effectiveness of the school's overall provision for pupils with SEND. The graduated approach offers the school, the pupil and his or her parents, a growing understanding of needs and of what provision the pupil requires to enable him or her to make good progress and secure good outcomes. Through successive cycles, the graduated approach draws on increasingly specialist expertise, assessments and approaches and more frequent reviews. This structured process gives teachers the information they need to match specific, evidenced-based interventions to pupils' individual needs.

Evidenced-based interventions

In recent years, a number of universities and other research organisations have produced evidence about the efficacy of a range of different interventions for vulnerable pupils and pupils with SEND. Most notable among this research is that sponsored by the Education Endowment Fund that offers schools valid data on the impact of interventions and the optimal conditions for their use. Other important sources of information about evidence-based interventions for specific areas of need are the Communication Trust 'What Works?' website and 'Interventions for Literacy' from the SpLD/Dyslexia Trust. Both sites offer transparent and clear information for professionals and parents to support joint decisions about provision.

The Equality Act 2010

Sitting alongside the Children and Families Act 2014, the requirements of the Equality Act 2010 remain firmly in place. This is especially important because many children and young people who have SEND may also have a disability under the Equality Act. The definition of disability in the Equality Act is that the child or young person has 'a physical or mental impairment which has a long-term and substantial adverse effect on a person's ability to carry out normal day-to-day activities'. 'Long-term' is defined as lasting or being likely to last for 'a year or more', and 'substantial' is defined as 'more than minor or trivial'. The definition includes sensory impairments such as those affecting sight or hearing, and, just as crucially for schools, children with long-term health conditions such as asthma, diabetes, epilepsy and cancer.

As the SEND Code of Practice (DfE 2015, p.16) states, the definition for disability provides a relatively low threshold, and includes many more children than schools may realise. Children and young people with some conditions do not necessarily have SEND, but there is often a significant overlap between disabled children and young people, and those with SEND. Where a disabled child or young person requires special educational provision they also will be covered by the SEND duties.

The Equality Act applies to all schools including academies and free schools, university technical colleges and studio schools; and also further education colleges and sixth form colleges – even where the school or college has no disabled pupils currently on roll. This is because the duties under the Equality Act are anticipatory in that they cover not only current pupils but also prospective ones. The expectation is that all schools will be reviewing accessibility continually and making reasonable adjustments in order to improve access for disabled pupils. When thinking about disabled access, the first thing that school leaders usually consider is physical access, such as wheelchair access, lifts and ramps. But physical access is only part of the requirement of the Equality Act and often is the simplest to improve. Your school's accessibility plan for disabled pupils must address all of three elements of planned improvements in access:

1. physical improvements to increase access to education and associated services;
2. improvements in access to the curriculum;
3. improvements in the provision of information for disabled pupils in a range of formats.

Improvements in access to the curriculum are often a harder nut to crack as they involve all departments and all teachers looking closely at their teaching and learning strategies and evaluating how effectively these meet the needs of disabled pupils. Often, relatively minor amendments to the curriculum or teaching approaches can lead to major improvements in access for disabled pupils and these often have a positive impact on the education of all pupils. For example, one school installed a Soundfield amplification system in a number of classrooms because a pupil with a hearing loss had joined the school. The following year, the cohort of Year 7 pupils had particularly poor speaking and listening skills and it was noticed that they were more engaged in learning when they were taught in the rooms with the Soundfield system. This led to improvements in progress for the whole cohort and significantly reduced the level of disruption and off-task behaviours in those classes.

Schools also have wider duties under the Equality Act to prevent discrimination, to promote equality of opportunity and to foster good relations. These duties should inform all aspects of school improvement planning from curriculum design through to anti-bullying policies and practice.

Significantly, a pupil's underachievement or behaviour difficulties might relate to an underlying physical or mental impairment which could be covered by the Equality Act. Each pupil is different and will respond to situations in his or her unique way so a disability should be considered in the context of the child as an individual. The 'social model' of disability sees the environment as the primary disabling factor, as opposed to the 'medical model' that focuses on the individual child's needs and difficulties. School activities and environments should be considered in the light of possible barriers to learning or participation.

Appendix 1.2 Departmental policy

Whether the practice in your school is to have separate SEND policies for each department or to embed the information on SEND in your whole-school inclusion or teaching and learning policies, the processes and information detailed below will still be relevant.

Good practice for pupils with SEND is good practice for all pupils, especially those who are 'vulnerable' to underachievement. Vulnerable groups may include looked-after children (LAC), pupils for whom English is an additional language (EAL), pupils from minority ethnic groups, young carers and pupils known to be eligible for free school meals/ Pupil Premium funding. Be especially aware of those pupils with SEND who face one or more additional vulnerabilities and for whom effective support might need to go beyond help in the classroom.

It is crucial that your departmental or faculty policy describes a strategy for meeting pupils' special educational needs within your particular curricular area. The policy should set the scene for any visitor, from supply staff to inspectors, and make a valuable contribution to the departmental handbook. The process of developing a department SEND policy offers the opportunity to clarify and evaluate current thinking and practice within the D&T team and to establish a consistent approach.

The SEND policy for your department is a significant document in terms of the leadership and management of your subject. The preparation and review of the policy should be led by a senior manager within the team because that person needs to have sufficient status to be able to influence subsequent practice and training across the department.

What should a departmental policy contain?

The starting points for your departmental SEND policy will be the whole-school SEND policy and the SEND Information Report that, under the Children and

Families Act 2014, all schools are required to publish. Each subject department's own policy should then 'flesh out' the detail in a way that describes how things will work in practice. Writing the policy needs to be much more than a paper exercise completed merely to satisfy the senior management team and Ofsted inspectors. Rather, it is an opportunity for your staff to come together as a team to create a framework for teaching D&T in a way that makes the subject accessible, not only to pupils with special educational needs and disabilities, but for all pupils in the school. It is also an ideal opportunity to discuss the impact of grouping on academic and social outcomes for pupils. Bear in mind that the Code of Practice includes a specific duty that 'schools must ensure that pupils with SEND engage in the activities of the school alongside pupils who do not have SEND' (DfE 6.2, p. 92).

We need to be careful in D&T that when grouping pupils, we are not bound solely by measures in reading and writing, but also take into account reasoning and oral language abilities. It is vital that social issues are also taken into account if pupils are to be able to learn effectively. Having a complement of pupils with good oral ability will lift the attitude and attainment of everybody within a group.

Who should be involved in developing our SEND policy?

The job of developing and reviewing your policy will be easier if tackled as a joint endeavour. Involve people who will be able to offer support and guidance such as:

- the school SEND governor;
- the SENCO or other school leader with responsibility for SEND;
- your support staff, including teaching assistants and technicians;
- the school data manager who will be able to offer information about the attainment and progress of different groups;
- outside experts from your local authority, academy chain or other schools
- parents of pupils with SEND;
- pupils themselves – both with and without SEND.

Bringing together a range of views and information will enable you to develop a policy that is compliant with the letter *and* principle of the legislation, that is relevant to the context of your school and that is useful in guiding practice and improving outcomes for all pupils.

The role of parents in developing your department SEND policy

As outlined in Appendix 1.1, Section 19 of the Children and Families Act 2014 raises the bar of expectations about how parents should be involved in and influence the work of schools. Not only is it best practice to involve parents of pupils with SEND in the development of policy, but it will also help in 'getting it right' for both pupils and staff. There are a number of ways, both formal and informal, to find out the views of parents to inform policy writing, including:

- a focus group;
- a coffee morning/drop-in;
- a questionnaire/online survey;
- a phone survey of a sample of parents.

Parents will often respond more readily if the request for feedback or invitation to attend a meeting comes from their son or daughter.

Where to start when writing a policy

An audit can act as a starting point for reviewing current policy on SEND or writing a new policy. This will involve gathering information and reviewing current practice with regard to pupils with SEND, and is best completed by the whole department, preferably with some input from the SENCO or another member of staff with responsibility for SEND within the school. An audit carried out by the whole department provides a valuable opportunity for professional development so long as it is seen as an exercise in sharing good practice and encourages joint planning. It may also facilitate your department's contribution to the school provision map. But before embarking on an audit, it is worth investing some time in a departmental meeting, or ideally a training day, to raise awareness of the legislation around special educational needs and disabilities and to establish a shared philosophy across your department.

The following headings may be useful when you are establishing your departmental policy:

General statement of compliance

- What is the overarching aim of the policy? What outcomes do you want to achieve for pupils with SEND?
- How are you complying with legislation and guidance?
- What does the school SEND Information Report say about teaching and learning and provision for pupils with SEND?

Example

All members of the department will ensure that the needs of all pupils with SEND are met, according to the aims of the school and its SEND policy . . .

Definition of SEND

What does SEND mean?

- What are the areas of need and the categories used in the Code of Practice?
- Are there any special implications for our subject area?

(See Chapter 1.)

Provision for staff within the department

- Who has responsibility for SEND within the department?
- What are the responsibilities of this role?

e.g.

- liaison between the department and the SENCO;
- monitoring the progress of and outcomes for pupils with SEND, e.g. identifying attainment gaps between pupils with SEND and their peers;
- attending any liaison meetings and providing feedback to colleagues;
- attending and contributing to training;
- maintaining departmental SEND information and records;
- representing the needs of pupils with SEND at departmental level;
- liaising with parents of pupils with SEND;
- gathering feedback from pupils with SEND on the impact of teaching and support strategies on their learning and well-being.

(This post can be seen as a valuable development opportunity for staff and the name of this person should be included in the policy. However, where responsibility for SEND is given to a relatively junior member of the team, there must be support and supervision from the head of the department to ensure that the needs of pupils with SEND have sufficient prominence in both policy and practice.)

- What information about pupils' SEND is held, where is it stored and how is it shared?
- How can staff access additional resources, information and training?
- How will staff ensure effective planning and communication between teachers and teaching assistants?
- What assessments are available for teachers in your department to support accurate identification of SEND?

Example

The member of staff with responsibility for overseeing the provision of SEND within the department will attend liaison meetings and subsequently give feedback to the other members of the department. S/he will maintain the department's SEND file, attend and/or organise appropriate training and disseminate this to all departmental staff. All information will be treated with confidentiality.

Provision for pupils with SEND

How are pupils' special educational needs identified?

e.g.

- observation in lessons;
- assessment of class work/homework;
- end of module tests/progress checks;
- annual examinations/ SATs/ GCSE;
- reports.

- How is progress measured for pupils with SEND?
- How do members of the department contribute to individual learning plans, meetings with parents and reviews?
- What criteria are used for organising teaching groups?
- How/when can pupils move between groups?
- What adjustments are made for pupils with special educational needs and/ or disabilities in lessons and homework?
- How do we use information about pupils' abilities in reading, writing, speaking and listening when planning lessons and homework?
- What alternative courses are available for pupils with SEND?
- What special arrangements are made for internal and external examinations?
- What guidance is available for working effectively with support staff?

Here is a good place also to put a statement about the school behaviour policy and any rewards and sanctions, and how the department will make any necessary adjustments to meet the needs of pupils with SEND.

Example

The staff in the D&T department will aim to support pupils with SEND to achieve the best possible outcomes. They will do this by supporting pupils to achieve their individual targets as specified in their individual learning plans, and will provide feedback for progress reviews. Pupils with SEND will be included in the departmental monitoring system used for all pupils.

Resources and learning materials

- Is any specialist equipment used in the department?
- How are differentiated resources developed? What criteria do we use (e.g. literacy levels)?
- Where are resources stored and are they accessible to both staff and pupils?

Example

The department will provide suitably differentiated materials and, where appropriate, specialist resources to meet the needs of pupils with SEND. Alternative courses and examinations will be made available where appropriate for individual pupils. Support staff will be provided with curriculum information in advance of lessons and will be involved in lesson planning. A list of resources is available in the department handbook.

Staff qualifications and continuing professional development (CPD)

- What qualifications and experience do the members of the department have?
- What training has already taken place, and when? What impact did that training have on teaching and learning, and progress for pupils with SEND?
- How is training planned? What criteria are used to identify training needs?
- How are SEND taken into account when new training opportunities are proposed?
- Is a record kept of training completed and ongoing training needs?

> **Example**
>
> A record of training undertaken, specialist skills and training required will be kept in the department handbook. Requests for training will be considered in line with the department and school improvement plan.

Monitoring and reviewing the policy

- How will the policy be monitored?
- Who will lead the monitoring?
- When will the policy be reviewed?

> **Example**
>
> The department SEND policy will be monitored by the head of department on a planned annual basis, with advice being sought from the SENCO as part of the three-yearly review process.

Conclusion

Creating a departmental SEND policy should be a developmental activity that will improve teaching and learning for all pupils, but especially for those who are vulnerable to underachievement. The policy should be a working document that will evolve and change over time; it is there to challenge current practice and to encourage improvement for both pupils and staff. If departmental staff work together to create the policy, they will have ownership of it; it will have true meaning and be effective in clarifying good practice.

An example of a departmental policy for you to amend is available in the eResources for this book at www.routledge.com/9781138714922

Appendix 3.1 Short-term planning checklist

Do your structures for short-term planning enable you to:

1. differentiate objectives from *schemes* of *work*

2. integrate targets from the *EHC plan*

3. detail *activities*

4. negotiate, agree and communicate about *roles* and *responsibilities*

5. use a range of *environments*, *resources* and *equipment*

6. *Group* learners in a variety of ways

7. develop balance and variety in approaches to *teaching* and *learning*

8. *Record* and *assess* learners' progress and achievement

. . . on a week-by-week, day-by-day, lesson-by-lesson basis?

Short-term planning should enable staff to:

• differentiate objectives from *schemes* of *work* to promote access, participation and achievement for all learners	• integrate targets from *EHC plans*, set in terms of key, cross-curricular skills, into curriculum-related group activities

(This should provide staff (and pupils/students) with clarity about **both** *curriculum-related learning opportunities* **and** *ways of addressing priorities for individual pupils/students within the group.)*

• detail *activities* which *all* learners will find challenging and accessible

(This should inform staff about what pupils/students, across a range of prior interests, aptitudes and achievements, are actually going to **do** *in lessons – with adaptations and extensions as appropriate.)*

• negotiate, agree and communicate about the *roles* and *responsibilities* of different members of the teaching team deployed in support of pupil/student learning

(This should provide everyone – teacher/lecturer, support staff, paramedics, volunteers etc. – with clarity about expectations of them and ways to check what they should be doing.)

• select, organise and use a range of *environments*, *resources* and *equipment* which is carefully matched to session content and priorities for individual pupils/students

(This should allow staff (and pupils/students) to prepare efficiently and effectively for each session of the day, providing for: changes of venue and equipment; and consistency of support where appropriate.)

• *group* students in a variety of ways matched to a variety of purposes

(This should encompass teaching: to the whole class; to large cross-class groups; in homogeneous 'sets'; in small mixed or 'jigsawed' groups; in pairs; or one-to-one etc.)

• develop balance and variety over time across a range of *teaching methods* and individual *approaches* to *learning* matched to lesson content and pupil/student preferences

(This should provide pupils/students with experience of, for example: active, investigative learning; watching and listening; problem solving; collaboration with peers; independent work; using ICT etc.)

• create opportunities to *record* and *assess* pupils'/students' progress and achievement in relation to **both** schemes of work **and** individual education/learning plans

(This should provide staff (and pupils/students) with manageable amounts of good quality evidence to support the processes of monitoring student development; reviewing targets; planning future teaching and learning; reporting progress and achievement – and monitoring school/college development.)

. . . on a week-by-week, day-by-day, session-by-session basis.

Appendix 3.2 Planning for differentiation

Session focus:

content	response

interest	structure

pace	teacher time

sequence	teaching style

level	learning style

access	grouping

Twelve kinds of differentiation

content

pupils work on various aspects of the same subject matter

interest

activities reflect pupils' own interests/experiences

pace

pupils work through material at varying speeds; work is presented at varying rates

sequence

pupils dip into material in varying orders – planned? self-selected?

level

pupils work on similar concepts at different levels, reflecting previous achievements

access

material is presented to pupils through varying modes – aural, visual, tactile, concrete, IT, symbols, linguistic etc.

response

pupils respond to similar activities in varying ways – may be planned (teacher requests varied outcomes) or spontaneous (pupils' responses vary)

structure

work presented in small, developmental steps or in conceptually-related chunks; subject-specific or integrated

teacher time

1:1 time with teacher; time allowed for responses; additional support time

teaching style

didactic? investigative? discursive?

learning style

listening? exploring? problem solving?

grouping

individual? pairs? groups? class? whole-school or department?

(*adapted from:* Lewis, A. (1992) 'From planning to practice', *British Journal of Special Education*, 19 (1), 24–27.)

Appendix 3.3 Time planner

Project

What have I got to do?

Week	*In Class*	*At Home*
1		
2		
3		
4		
5		
6		

- Research and design ideas
- Design proposal
- Planning and making
- Testing and modifying
- Making your final product
- Presenting your work
- Evaluating

(from *Technopacks*)

Date	Title	Name

stick in your design words here

Are you happy with your design?

:) :| :(

Y

?

N

(from *Technopacks*)

Appendix 3.4 Key words

Ingredient words

Action words

(from *Technopacks*)

Equipment words

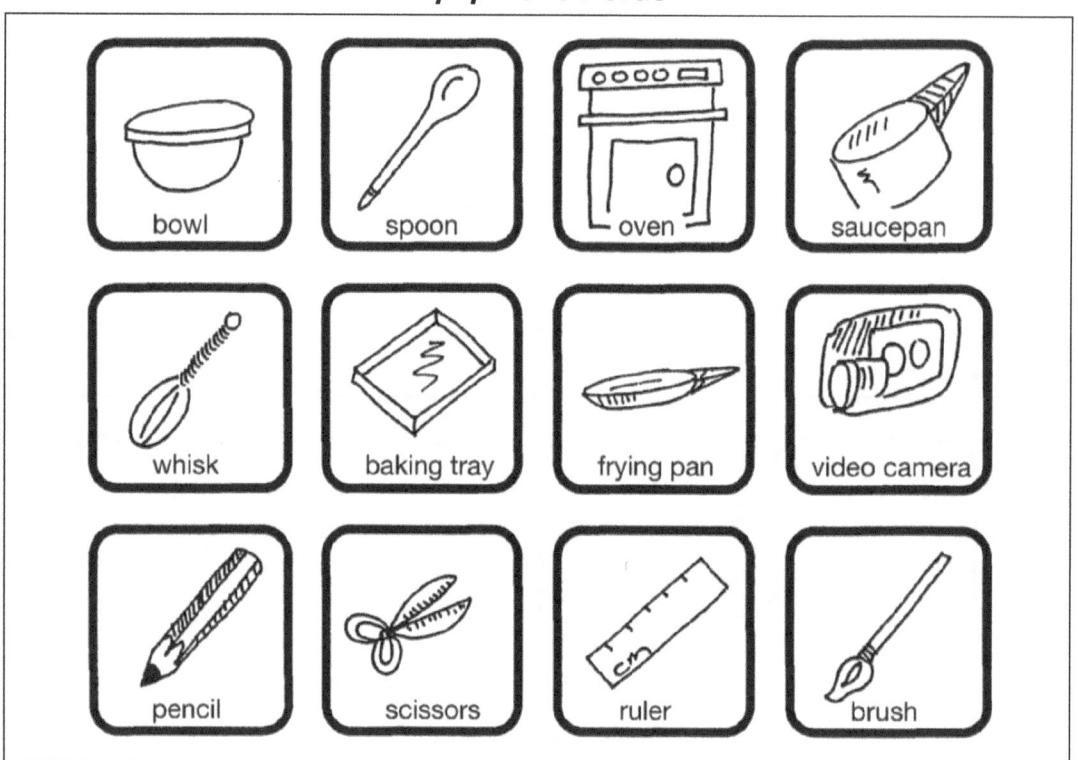

Material words

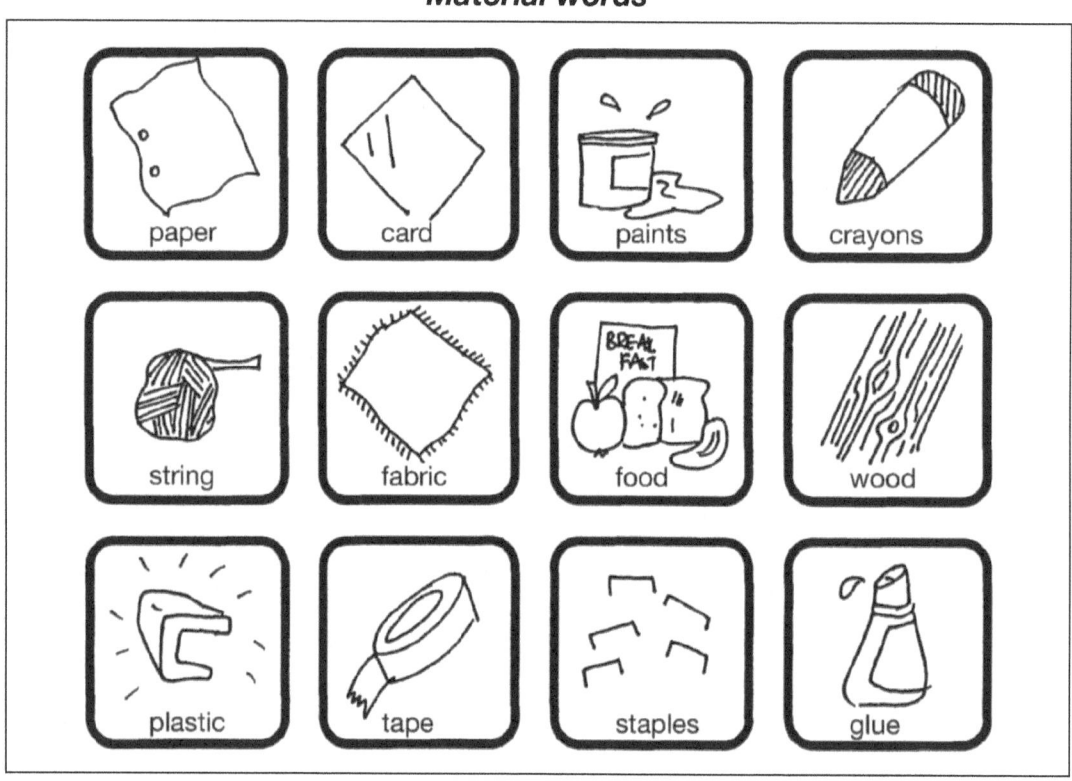

(from *Technopacks*)

Appendix 3.5 SEND design sheet

Use the pictures to help your pupils take part in all activities, especially design. They are divided into three sections: action, equipment and ingredient or material words.

Cut out the 'word pictures' which are the most suitable for your pupil.

Glue these on to the main design sheet.

design area   design words

Point to the pictures, like Bliss, running your finger down until the pupil indicates the word they want.

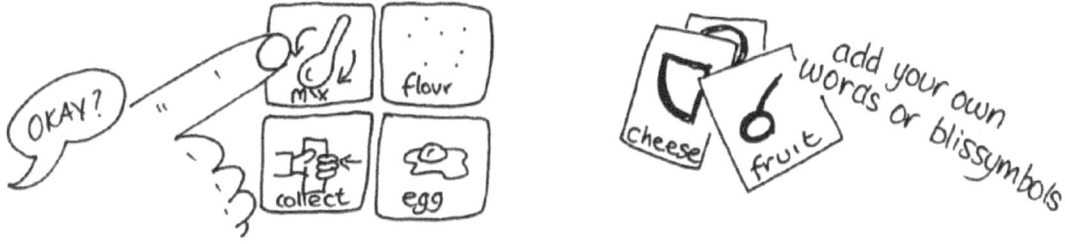

Alternatively they could be cut out and used by the pupils to show what equipment/action/ingredient they have used. These could then be glued into their work books. (Helpful for ESL pupils.)

The design, recipe or other activity can therefore be built up through questioning the pupil. With you putting the pupil's ideas into the centre of the design area.

Activities with a minimum of stages may be ideal as they cut down on the amount of processes involved, plus the amount of equipment and ingredients/materials.

Design picture words

Action words

(from *Technopacks*)

Design picture words

Your own words

<table>
<tr><td></td><td></td><td></td><td></td></tr>
<tr><td></td><td></td><td></td><td></td></tr>
<tr><td></td><td></td><td></td><td></td></tr>
</table>

(from *Technopacks*)

Appendix 3.6 Design ideas planner

Design ideas planner 1

What is the problem?

Who is your target group? (Who will your design be for?)

What are the constraints or limits on your designing and making?

Initial ideas to
be explored

Design ideas planner 2

Sketch 3 possible design ideas

- Where did you get these ideas from?
- How will these ideas help the problem?
- What are the good and bad points of each design?
- Who are your designs for?
- How will your designs be made?
- Will you be able to make them within the constraints?

Choose your best design and explain why it is the best

(from *Technopacks*)

Appendix 3.7 Exemplar planned sequence of questions

The group is working on their last lesson of a D&T unit and is designing and making a pizza. The class comprises a small group of Key Stage 3 pupils with a broad range of learning difficulties, including pupils with multiple sensory and physical difficulties. The teacher wants to engage the pupils in analysis and evaluation, though at a fairly concrete level.

- What have we been doing? (**Knowledge** – recall)
- What types of pizza have we made? (**Knowledge** – recall)
- What toppings did we use? (**Knowledge** – recall)
- *What was your favourite?* (**Knowledge** – recognise/identify/select)
- *What was Jason's favourite?* (**Knowledge** – recognise/identify/select)
- *What is this called?* (**Knowledge** – identify)
- Is it bacon or a green pepper? (**Knowledge** – identify)
- What do we do before we start making a pizza? Why? (**Comprehension** – describe and explain)
- What are we going to make with flour and margarine? (**Application**)
- How do we mix them together? (**Comprehension** – describe)
- How do we make the dough fit in the tray? (**Application** – solve)
- What do we put on the pizza next? (**Analysis**)
- *Then what will we put on?* (**Application** – sequence)
- *Is the pizza spicy?* (**Knowledge** – identify)
- *Does it taste salty or sweet?* (**Analysis** – compare/identify)
- Do you like sweet pizza or savoury pizza? (**Analysis** – compare/identify/select)
- What would happen if we added pineapple? Would it make it salty or sweet? (**Synthesis** – hypothesise)
- What topping would you leave out if you made it again? Why? (**Evaluation** – evaluate)

(SEN: Training materials for the foundation subjects,
DfE Publications, 2003)

Appendix 3.8 Plenary case study

Subject	Design and technology
Topic	Food technology – 'Designing and making a pizza'
Teaching context	Mixed ability Years 8 and 9 class. Learning difficulties range from severe to profound and multiple. There are several pupils with autistic spectrum disorders.

Having tasted and evaluated a range of pizzas during the lesson, and considered lists of possible pizza toppings in a previous lesson, the pupils return to the group for the plenary. They are informed that, in the next lesson, they will make a giant pizza.

A large pizza plate is placed in the middle of the table with pictures of the toppings that have been evaluated around it on the table. The pupils are asked to add the ingredients to the plate in the correct sequence for making a pizza. For example: 'What goes onto the plate first?', 'the base', 'then what comes next?', 'tomato purée, cheese', and so on. Each pupil is encouraged to select a topping to add to the pizza. The responses expected will vary according to a given pupil's communication mode and may include saying the words, signing, using a communication aid or eye-pointing between the items. Throughout the process the teacher prompts pupils, where appropriate, to give reasons for their choices. The activity encourages pupils to understand the link between the learning that has just taken place and the next lesson.

The group then reviews the choices made with the option to change or omit a topping that they collectively dislike or does not 'go' with the others, for example marshmallows on a savoury pizza. The symbols are then removed from the plate and glued onto a shopping list of ingredients to be purchased for the following lesson.

(*SEN: Training materials for the foundation subjects*,
DfE Publications, 2003)

Appendix 3.9 Pre-course task: Modelling

Definition of modelling

Modelling involves the teacher as 'an expert' demonstrating how to do something while thinking through the process aloud. By thinking aloud the teacher shows the importance of making decisions such as:

how to begin the task;
how to select information which is relevant to the task or audience;
how to organise the information or ideas;
the use of protocols relating to the presentation of information or ideas;
how to end the task.

Examples of modelling

Processes, concepts or skills which could be modelled include:

writing an account or explanation;
constructing a mind-map;
evaluating a piece of artwork or a finished product in D&T;
considering options when receiving the ball in an invasion game, for example football or netballl;
making a fruit kebab.

The following is an extract from a teacher commentary which illustrates the last example.

Teacher (modelling at a table on which are arranged a range of different fruits, kebab sticks, peelers and knives)

Has anyone ever had or seen a kebab?

I want to make a fruit kebab.

What do I need to do first?

Have I got all my ingredients?

Which fruit might go well together?

I like melon, strawberry and kiwi fruit.

I have chosen these because I like the taste of them. They have different colours and textures. I also think I could make a pattern on my stick with these fruits.

I'm going to choose my fruit now.

At this point, the pupils can be asked, 'What fruit are you going to choose?' They can then model back the process of choosing their fruit, using appropriate modes of communication.

What do I need to do next?

I am going to prepare/peel my fruit.

How big do I want the pieces to be? They have to fit on a kebab stick, so they have to be fairly big. They must not be too big because I want to fit them into my mouth.

What shape would be best? I'm going to cut mine into cubes because it is easy to cut cube shapes and they would look good on a stick.

At this point, the pupils can be asked about the decisions they have made about the size and shape of their pieces of fruit. Again, they can model back their thinking.

If any pupils have difficulty in cutting the fruit, the teacher can put the pupil's hand on his/her hand to guide them through the cutting motion.

How many pieces am I going to need? I'm going to look at the length of the kebab stick and estimate.

Now I'm going to put my fruit on the stick.

Which one am I going to start with? I'm going to start with a piece of melon because it is firm and will keep the other pieces of fruit on the stick.

Which piece am I going to put on next?

No, I don't like that one (discards pineapple), because it is too similar to the colour of the melon. I'm going to put a red one on, then a green one.

Here, the pupils can discuss the decisions they have made about the order of their fruit and the pattern they are going to make.

Pupils can be encouraged to reflect on the process.

Which fruit did I use? Why?

Does it look tasty/appetising?

Are there any changes I could make to improve the kebab?' (think of taste, colour, pattern, texture, combination).

Task

In preparation for this module try to model a new skill, concept or process to one of your classes.

Reflect on the success of the modelling strategies you used and note down your responses to the following three questions:

1. How effective were your modelling strategies?
2. What did you do to help pupils to apply the strategies you modelled independently, for example prompts, practice and discussion time?
3. What might you do differently next time?

These questions will be discussed in the training session for module 6 'Modelling'.

The module will also look at how pupils can be helped to apply skills and strategies independently.

(SEN: Training materials for the foundation subjects,
DfE Publications, 2003)

Appendix 3.10 Self-audit for an inclusive D&T classroom

These tables are designed to help you to reflect upon aspects of D&T teaching that can improve the learning of pupils with SEND. However, many of the ideas will help all pupils.

The physical environment – general	Design and technology	Observed	Tried-out
Visual prompts for routines, such as how to ask for help. Visual timetables showing plans for the day or lesson.	Visual prompts using pictures, photos or symbols prepared for the order in which a sequence of activities is to be carried out.		
Background noise avoided. Sound and light issues considered. Video/TV positioned so that all pupils can see/hear.	Height adjustable tables, sinks and hobs aid pupils to access activities. Avoid demonstrations or discussion when machines are running. Film demonstrations will require subtitles for deaf or hearing-impaired pupils.		
Seating should allow for peer or adult support; space between chairs for pupils with mobility difficulties; clear view of the board for all pupils; and room for left-handers. Need to consider whether pupils can see/hear the teacher and see the board clearly.	Plan your demonstration area so that it is clearly laid out, uncluttered and gives all pupils a clear view. Plan how to introduce new skills to the class (and individuals) and ensure that TAs are familiar with how to demonstrate new skills correctly.		
Resources are accessible and clearly labelled. Equipment colour coded and labelled to encourage independent use helps everyone.	Labelling cupboards with a digital photo/name label helps in deaf access to equipment.		

Planning for the lesson – general	Design and technology	Observed	Tried-out
Draw on the breadth of curriculum possibilities in order to involve all pupils equally in tasks.	When a pupil only wants to 'make', choose a task that will only work if some designing is done too. Look for tasks that motivate individuals and are relevant to their lives. Pupils with ASD may have low awareness of danger at any age. Visits to supermarkets, cafes and DIY stores allow pupils to experience shopping, traffic and weather changes, which furthers their understanding of the world around them.		
Support planned for individuals or groups with SEND in terms of differentiated resources, e.g. large-font handouts, simpler (or extension) tasks.	Use jigs, templates, patterns, pre-cut or pre-made parts if coordination and accuracy are issues.		
Concepts and language taught through multi-sensory and cross-curricular approaches.	Pictures, signs, symbols, computer animation and audio recording can all help.		
Questions prepared in different styles/levels for different pupils	Use simple choice cards with words and symbols such as like/dislike for product evaluation. Simple ranking and sensory-testing recording sheets, using symbols and words to point to or circle can record individual responses		
A distraction-free (low arousal) area planned for pupils who may need it and available for use by all pupils.	Checklists and visual support allow pupils to see what has been completed, what to do next and where to end.		
The next stages of learning are signposted and highlighted for individuals or groups.	Segment the designing and making stages into small manageable steps, and incorporate designing into 'mini-making' tasks.		
Plan for 'scaffolding', the process of identifying what a pupil can do with support, and then taking the support away in stages, until the pupil can work independently.	TAs can promote independence through questions that extend pupil learning and guidance, rather than doing this task for them.		
Other adult support targeted at individuals or groups.	This should include specific support with learning health and safety rules.		
TAs clear about learning objectives/individual targets and deployed to provide only the support pupils need, whilst also encouraging pupils to work independently when they can	TAs are clear about the skills and knowledge they must promote, such as basic food hygiene.		
Plan to explain independent study activities/homework *during* the lesson so pupils have plenty of time to record them and/or access to copies of the task.			

During the lesson – general	Design and technology	Observed	Tried-out
All pupils clear about the duration and overall structure of the day and the lesson. Visual timetables have been referred to.			
All pupils clear about the planned outcomes at the start of the lesson.	Give an overview with a few straightforward instructions.		
Key words, meanings and symbols highlighted and explained.	Clarify technical terms that have meaning in other contexts, such as: knead/need, grain, glaze, form, saw, seam etc.		
Recognise pupils' responses to errors, value them and build the on the though processes involved for the pupils.	Show how mistakes can be put right to remove the fear of making mistakes.		
Paired talk or buddy talk is encouraged to maintain attention or to link concepts to pupils' own varied experiences			
Manageable mixed attainment grouping or pairing is the norm, except when carefully planned.			
The transition from whole-class work to independent/group work, and back, is clearly signalled.			
Oral interactions and explanations are valued.	Balance asking the right questions against talking so much that it prevents pupils moving forward. Many pupils find talking about what they have done easier than talking about their plans.		
Alternatives to written recording offered, where appropriate (e.g. drawing, scribing, mind maps, ICT-based approaches, including CAD/CAM).	Using digital cameras at each stage of designing and making, then sequencing them, is a useful tool to aid memory of work completed. Encourage modelling with scrap materials.		

The end of the lesson and afterwards – general	Design and technology	Observed	Tried-out
End-of-lesson discussion focuses on one or more of the ideas/skills explored and the progression towards them during the lesson.	Make sure that pupils, particularly those with BESD, judge work against the design specification work rather than that of other pupils.		
End-of-lesson discussion focuses on the ways of working the class have found fruitful.			
Main points from the sequence of lessons fed back by pupils, including inputs from pupils with SEND, noted down and displayed where everyone can see them.	Use digital camera or video to record experiences and responses and show them when feeding back to pupils.		

Appendix 4.1 Long-term planning sheets

Long-term planning sheet

	Design and make assignment		*Design and make assignment*		*Design and make assignment*	
	How things work		*Ourselves (masks)*		*A card for all occasions*	
Year	**UNIT**	TIME	**UNIT**	TIME	**UNIT**	TIME
Cycle	**A**		**B**		**C**	
ABC	**Designing and making foci**		**Designing and making foci**		**Designing and making foci**	
	Working with materials		Clarifying the task		Clarifying the task	
	Generating ideas		Communicating intentions		Communicating ideas	
	Health and safety		Developing ideas		Working with materials	
	Knowledge and understanding		**Knowledge and understanding**		**Knowledge and Understanding**	
	Pulleys and gears		Sheet materials		Sheet materials	
	Wheels and axles		Mouldable materials		Reclaimed materials	
	Electrics		Textiles		ICT	
			Products and applications			
	Design and make assignment		*Design and make assignment*		*Design and make assignment*	
	Homes		*Hand warmer*		*Fairground games*	
Year	**UNIT**	TIME	**UNIT**	TIME	**UNIT**	TIME
Cycle	**D**		**E**		**F**	
ABC	**Designing and making foci**		**Designing and making foci**		**Designing and making foci**	
	Clarifying the task		Clarifying the task		Clarifying the task	
	Working with materials		Developing ideas		Communicating intentions	
	Evaluating		Evaluating		Evaluating	
	Knowledge and understanding		**Knowledge and understanding**		**Knowledge and understanding**	
	Structures		Textile materials		Sheet materials	
	Reclaimed materials		Products and applications		Framework materials	
	Sheet materials				Structures	
					Electrics	

Design and make assignment	*Design and make assignment*	*Design and make assignment*
Picnics	*Flip-flops at the seashore (slipper)*	*Night light*

	UNIT TIME	**UNIT** TIME	**UNIT** TIME
Year	**G**	**H**	**I**
Cycle	**Designing and making foci**	**Designing and making foci**	**Designing and making foci**
ABC	Developing ideas	Clarifying the task	Clarifying the task
	Planning	Generating ideas	Working with materials
	Health and Safety	Evaluating	Health and safety
	Knowledge and understanding	**Knowledge and understanding**	**Knowledge and understanding**
	Food materials	Mouldable materials	Electrics
	Sheet materials	Products and applications	Sheet materials
			Framework materials

Long-term planning sheet

	Autumn term	*Spring term*	*Summer term*
	Design and make assignment	*Design and make assignment*	*Design and make assignment*
	Puppets	*Pizzas (link picnics)*	*Boats*

	UNIT TIME	**UNIT** TIME	**UNIT** TIME
Cycle 1	**J**	**K**	**L**
ABC	**Designing and making foci**	**Designing and making foci**	**Designing and making foci**
	Clarifying the task (knowing what to do)	Clarifying the task	Clarifying the task
	Developing ideas (response to different designs)	Generating ideas	Working with materials
		Evaluating	Health and Safety
	Working with materials (and response to them)	Health and Safety	**Knowledge and understanding**
	Knowledge and understanding	**Knowledge and understanding**	Reclaimed materials
	Frameworks	Food materials	Textiles
	Textiles		Framework materials
	Structure		

	Design and make assignment	*Design and make assignment*	*Design and make assignment*
	Bridges	*Vehicles – on the move*	*Pop-up big book page*

	UNIT TIME	**UNIT** TIME	**UNIT** TIME
Cycle 2	**M**	**N**	**O**
ABC	**Designing and making foci**	**Designing and making foci**	**Designing and making foci**
	Generating ideas	Clarifying the task	Developing ideas
	Planning (drawing)	Generating ideas	Planning
	Evaluating	Working with materials	Communicating intentions

Knowledge and understanding Structures Framework materials Sheet materials	Health and safety **Knowledge and understanding** Wheels and axles Reclaimed materials Products and applications	**Knowledge and understanding** ICT Sheet materials Levers and linkages
Design and make assignment *Fairgrounds (roundabouts)*	*Design and make Assignment* *Bird feeder/table*	*Design and make Assignment* *Party (table mats)*

	UNIT TIME	**UNIT** TIME	**UNIT** TIME
Cycle 3 *ABC*	**P** **Designing and making foci** Working with materials Generating ideas **Knowledge and understanding** Cans and cranks Electrics Pulleys and gears	**Q** **Designing and making foci** Clarifying the task Health and safety Generating ideas Evaluation **Knowledge and understanding** Structures Food Sheet material	**R** **Designing and making foci** Communicating intentions Planning Evaluating **Knowledge and understanding** Food Sheet materials ICT

Long-term planning sheet

	Autumn term	*Spring term*	*Summer term*
	Design and make assignment *Shelter – fabric structures*	*Design and make assignment* *Pop-up surprise*	*Design and make assignment* *Bird food (link bird feeder)*
Year *Cycle* *ABC*	**UNIT** TIME **S** **Designing and making foci** Clarifying the task Generating ideas Evaluating **Knowledge and understanding** Textile materials Framework materials Structures	**UNIT** TIME **T** **Designing and making foci** Generating ideas Developing ideas Working with materials **Knowledge and understanding** Pneumatic Hydraulic Reclaimed materials	**UNIT** TIME **U** **Designing and making foci** Clarifying the task Generating ideas Evaluating **Knowledge and understanding** Food materials
	Design and make assignment	*Design and make assignment*	*Design and make assignment*
Year *Cycle* *ABC*	**UNIT** TIME **Designing and making foci** **Knowledge and understanding**	**UNIT** TIME **Designing and making foci** **Knowledge and understanding**	**UNIT** TIME **Designing and making foci** **Knowledge and understanding**

	Design and make assignment	*Design and make assignment*	*Design and make assignment*
Year	**UNIT** TIME	**UNIT** TIME	**UNIT** TIME
Cycle ABC	**Designing and making foci**	**Designing and making foci**	**Designing and making foci**
	Knowledge and understanding	**Knowledge and understanding**	**Knowledge and understanding**

Whole-school long-term planning sheet (blank)

	Autumn term	*Spring term*	*Summer term*
	Design and make assignment	*Design and make assignment*	*Design and make assignment*
Cycle 1 ABC	**UNIT** TIME	**UNIT** TIME	**UNIT** TIME
	Designing and making foci	**Designing and making foci**	**Designing and making foci**
	Knowledge and understanding	**Knowledge and understanding**	**Knowledge and understanding**
	Design and make assignment	*Design and make assignment*	*Design and make assignment*
Cycle 2 ABC	**UNIT** TIME	**UNIT** TIME	**UNIT** TIME
	Designing and making foci	**Designing and making foci**	**Designing and making foci**
	Knowledge and understanding	**Knowledge and understanding**	**Knowledge and understanding**
	Design and make assignment	*Design and make assignment*	*Design and make assignment*
Cycle 3 ABC	**UNIT** TIME	**UNIT** TIME	**UNIT** TIME
	Designing and making foci	**Designing and making foci**	**Designing and making foci**
	Knowledge and understanding	**Knowledge and understanding**	**Knowledge and understanding**

Appendix 4.2 Differentiating lessons

Subject: Design and technology   Module: Using control to control a display   Class/group: Key Stage 3, Group 2

Sample lesson plan

Learning objectives	Learning outcomes	Content, activities and strategies
Group 1 – Ben, Sinead, Jamie *Be able to:* Name the components in an electrical circuit; Know how a simple circuit works; Design and make (Ben with TA support) a simple electrical circuit on your display board. **Group 2 – Carl, James, Jenny** *Be able to:* Name the components in a mechanism you have designed; Know how levers and pulleys work; Make moving parts on your display board using levers (Carl and Jenny with TA support). **Group 3 – Daniel, Becky, Michael** *Be able to:* Know what a graphic is; Create and change graphic lettering; Choose appropriate lettering for your graphic; Make your graphic 3D on your display board.	Describe the electrical components in their product and explain what they do; Design and make a simple electrical control circuit (e.g. Ben – spinning football, Sinead – Barbie light, Jamie – spinning sun). Describe the mechanical components in your product and explain what they do; Name the equipment you used; Make a mechanism for the motion you chose (e.g. Carl – jumping dog, James – waving hand, Jenny – tongue moving in/out). Describe your graphic's features; Create and change the lettering; Improve the appearance of your product (e.g. Daniel – 2D/3D, Becky – font size and colour, Michael – font size and colour).	**Starter** Whole class: Draw a mind map to recap previous learning. Set individual pupil objectives and explain the structure of the lesson. Explain to pupils what is expected of them and what they will accomplish by the end of the lesson. Use pupils' display boards to demonstrate expectations for today's work on graphics, electrical and mechanical control. Use pictures, symbols, signs and modified language to communicate this. Four TAs should support learning as previously agreed. **Main part** Group 1 (TA: Anne): Electrical circuits, using components to make an electrical part for their board. Group 2 (TA: Amanda): Making levers and other mechanical moving parts to go on their boards, use of craft knives, paper fasteners, etc. Group 3 (TA: Jacquie): Pictures and graphics using the computers, 2D and 3D, use of the laminator. Teacher to work with all groups to reinforce objectives and give feedback. Cue pupils and staff regarding timing of the lesson and the end of the lesson. **Plenary** Whole class: recap focus of each group's work. Ask questions to determine whether learning objectives/outcomes have been achieved. Ask pupils to evaluate their own and their peers' work. Agree next steps for individuals.
Key vocabulary: display, product, control, movement, mechanical, electrical, design, make, names of materials, construction, construction methods.	**Resources:** electrical circuits, card, corroflute, fabrics, camera, photos, Word art, clip art, glue, scissors, tape, double-sided adhesive tape, masking tape, paper fasteners, craft knives, string, pulleys; additional materials as required.	

(Adapted from: DfE, *SEN: Training materials for the foundation subjects* (DfE Publications, 2003)

Appendix 4.3 Using the project pro forma

Each project pro forma has various sections illustrated below. A blank pro forma follows, should you wish to use the planning format.

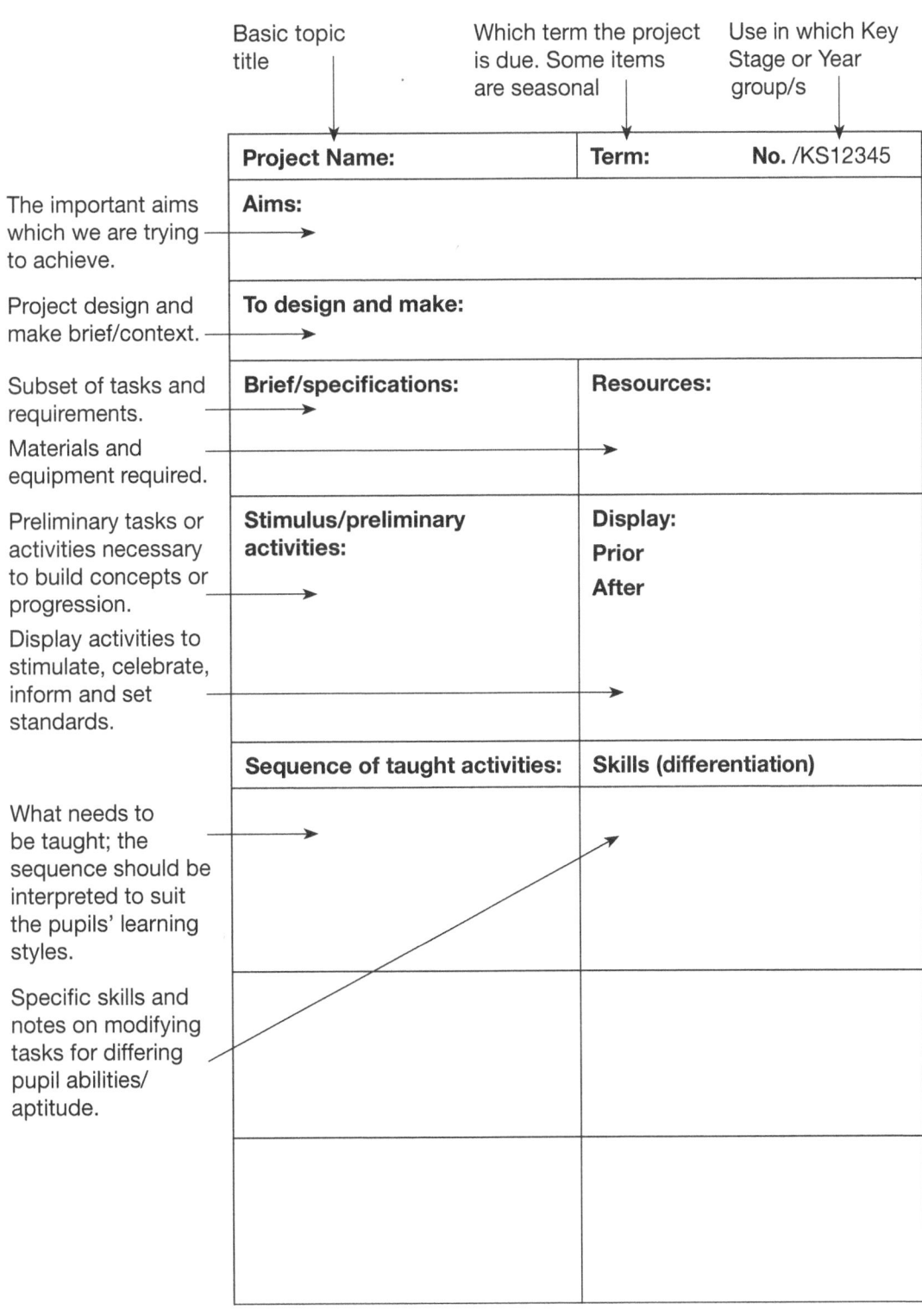

Project name:	Term: No.
Aims:	
To design and make:	
Brief/specifications:	**Resources:**
Stimulus/preliminary activities:	**Display:** **Prior** **After**
Sequence of taught activities:	**Skills (differentiation)**

Appendix 4.4 Key words

Ingredient words

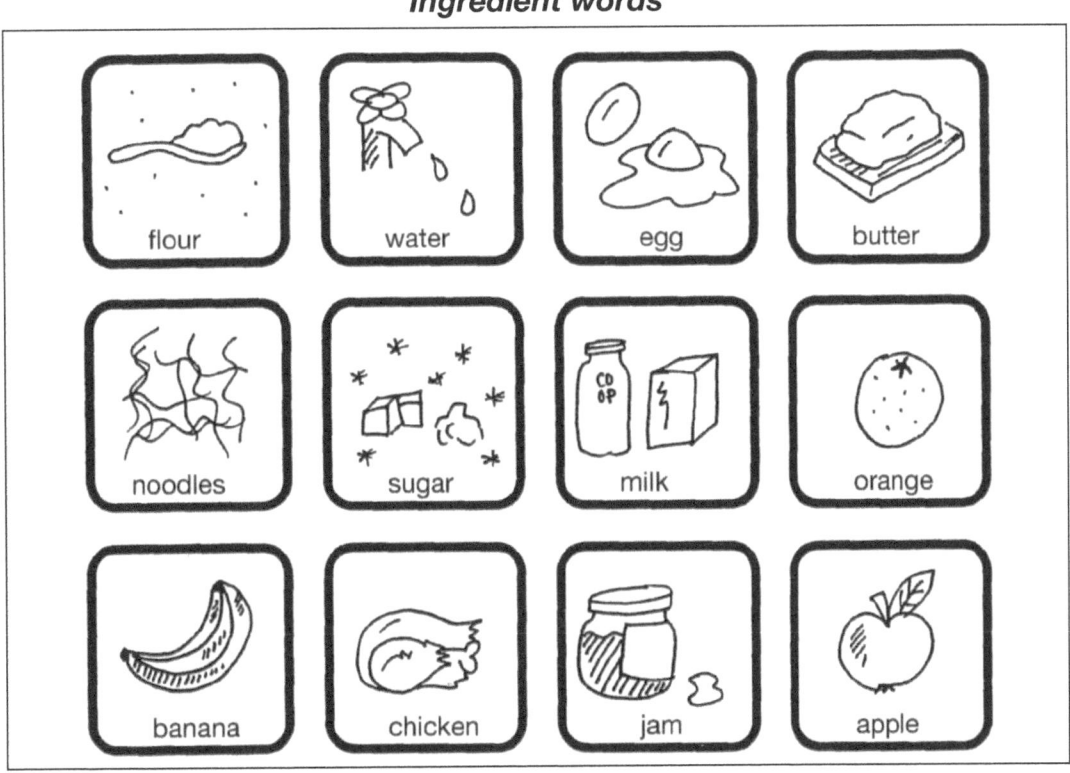

Action words

(from *Technopacks*)

Equipment words

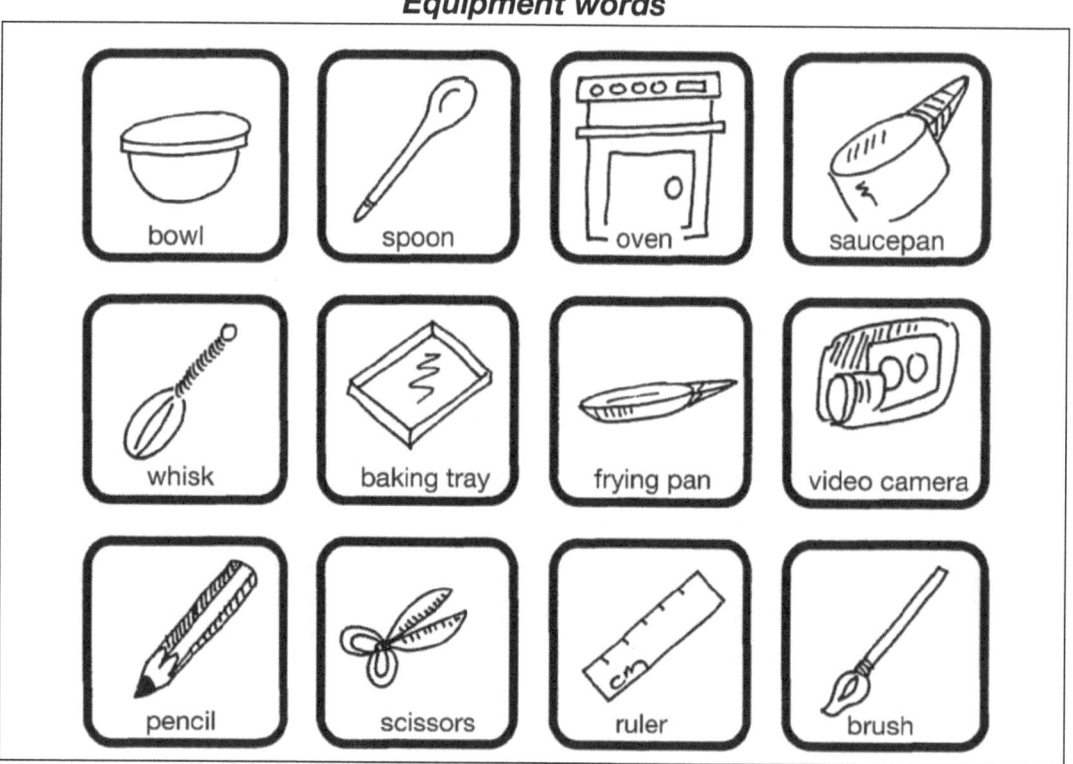

Material words

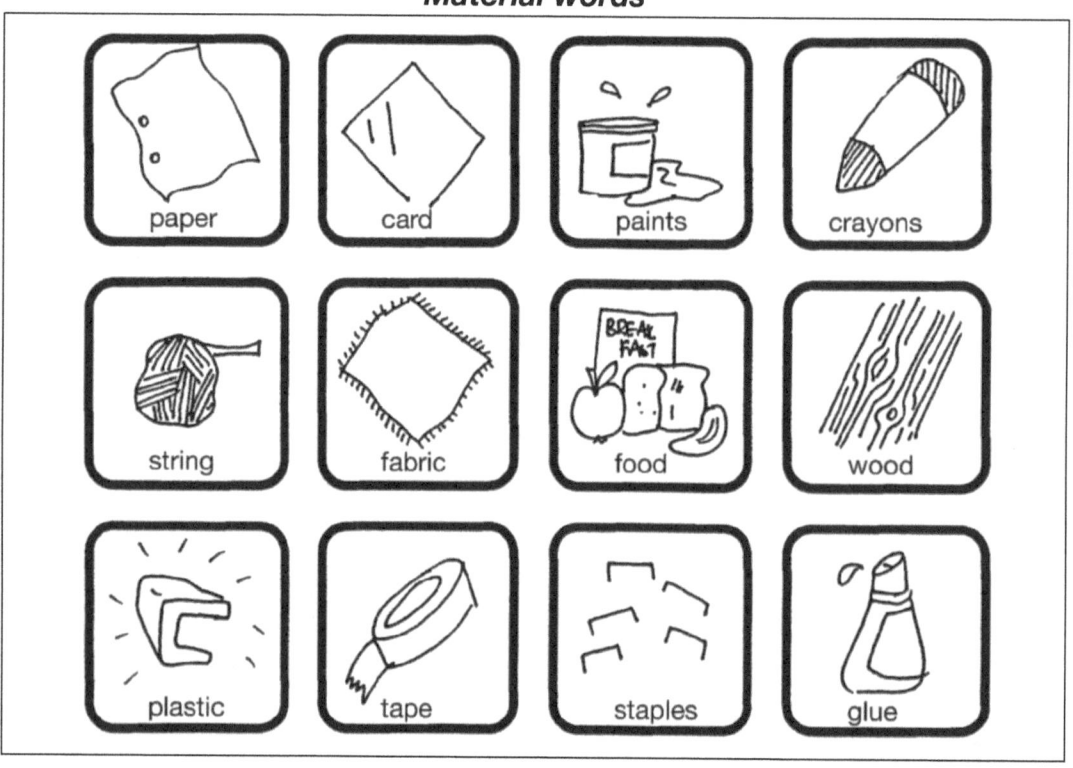

(from *Technopacks*)

Appendix 6.1 Differentiation and progression steps: Puppets project

Generating ideas	Developing ideas	Textiles	Quality
Considering possibilities. Evolving/Diversifying ideas.	Modelling/refining/specifying ideas.	Fabrics, yarns and threads, components (e.g. buttons).	Fitness for purpose. Construction methods, material choice.
Formative D&T skills and knowledge – Pupils should be able to:			
Suggest what might be done through the exploration of a collection of familiar products with the teacher (products and applications). **Use** the knowledge gained to suggest ways of making similar items. **Explore** ideas by rearranging materials.	**Use** reclaimed materials to 'model' their thinking. **Select** pictures that help the development of ideas.	**Classify** fabrics by texture, colour and finish. **Use** fabric crayons to apply decoration. **Cut** and stick fabrics (e.g. dressing a cardboard figure).	**Examine** and appreciate that products are made from different materials. **Say** what they do and do not like about products. **Collect** similar products.
Next steps 1 – In addition, pupils may be able to:			
Use their knowledge of an existing product to inform ideas for similar products of their own (e.g. gloves or boxes). **Propose** more than one idea for the same product.	**Use** a combination of kits and reclaimed materials to 'model' their thinking. **Use** kits to develop more than one idea and use them to refine the criteria for the final product. **Use** drawings to record ideas as they are developed. **Add** notes to drawings to help explanations. **Discuss** their work as it progresses.	**Cut** and join fabrics by using running stitch, glue, staples, buttons, Bondaweb etc. **Decorate** fabric by painting, printing, applying beads and sequins etc. **Use** paper patterns.	**Describe** how a product is made, how many parts it is made from, how they are joined, and how movement is created by mechanisms. **Identify** the different materials that are used for different parts of a product. **Name** some of the materials. **Seek** views about a product from users.

Next steps 2 – In addition, pupils may be able to:

Use the investigation of similar products as the one to be made to give starting points for the design.

Use drawings to help analyse and understand how products are made and to inform thinking about their own product.

Propose realistic suggestions as to how they can achieve their design ideas.

Use a combination of modelling and drawing to refine ideas about the function.

Make decisions about what will make a successful product.

Recognise basic properties – water resistance, texture, strength, etc.

Recognise the faces of fabrics and the need for seams.

Apply patterns using a variety of embroidery stitches.

Describe the way in which a product has been made.

Discuss and decide what the design criteria might have been for a product.

Evaluate how well a product meets the design criteria.

Next steps 3 – Pupils may make further progress if they can:

Use knowledge of materials and systems to inform their thinking about their design.

Use models, kits and drawings to help formulate their design idea.

Suggest a variety of ways in which one idea might be made, with modifications for improvement.

Use modelling and drawing skills to develop a specific idea in some depth.

Use information found through drawing and modelling to inform decisions about the design.

Use back-stitch and tacking.

Hem and reinforce edges and seams.

Combine fabrics to create more useful properties.

(e.g. interfacing to stiffen or wadding to insulate).

Draw conclusions from what users say about a product.

Suggest alternative designs and ways of making a similar product.

Consider alternative materials and constructions.

Consider old and new products and compare them.

Appendix 6.2 Are your targets S.M.A.R.T?

Specific

- are the targets specific to this particular pupil?
- do they address 'extra' or 'additional' priorities and avoid unnecessary duplication of objectives already specified in schemes of work?

Measurable

- do the targets address particular skills or accomplishments?
- do they specify proposed levels of prompting or support?
- are they set in terms of well-defined, observable outcomes so that:

  - you are looking for significant, new responses?
  - everyone knows what to record and when?
  - staff will know when the target is achieved?
  - the pupil will know when the target is achieved?

Achievable

- is the progress implied by the targets realistic:

  - for this pupil?
  - in this time-frame?
  - with these levels of resourcing?
  - should the targets be broken down into smaller steps?

Relevant

- do the targets address the priority areas of learning for this pupil?
- are they designed to meet real needs in the whole curriculum?
- is there a small number of tightly-focused targets?
- will everyone involved understand the circumstances or context in which the target can be addressed?

Time-limited

- when is the agreed short-term review for these targets?
- should the timescale be more tightly focused?
- if the target is achieved within the timescale, what next?
- It is worth noting that some targets may involve:

  - experimentation and exploration in new areas of learning;
  - consolidation, maintenance, transfer, generalisation of pre-existing skills, knowledge, understanding, etc;
  - slowing or reduction in rates of regression.

A 'method' section aligned to targets should be used to detail the circumstances or contexts in which the targets can be addressed – activities, resources, equipment, staff roles, pupil groupings, rewards, etc.

Individual support programmes can be used to give information about consistent approaches to the management of physical positioning, mobility, aided and augmented communication, problematic behaviour, sensory impairments, personal care regimes, medical and paramedical issues, therapies and therapeutic approaches, counselling, etc.

Appendix 6.3 Prompt sheets

Name _____

		I worked alone	I worked as part of a group	I needed help
	LOOKING AT LABELS			
	SORTING OUT IDEAS – concept screening			
	WRITING A NEW FILLING – specification			
	PROTOTYPE 1			
	CONSUMER TESTING 1			

Name _____

	I worked alone	I worked as part of a group	I needed help
EVALUATION 1			
PROTOTYPE 2			
CONSUMER TESTING 2			
EVALUATION 2			
COSTING			
FINAL EVALUATION			

Appendix 6.4 Useful stems for writing objectives

- **Show awareness of . . .** (through gestures, signs, symbols or words).
- **Respond to . . .** (through gestures, signs, symbols or words).
- **Recognise . . .**
- **Communicate about . . .**
- **Know . . .** (recall).
- **Be able to . . .** (apply skills or processes).
- **Understand . . .**
- **Develop** *or* **Are aware of . . .** (attitudes and values).

Appendix 6.5 Sample lesson plan

Subject: Art and design
Topic: Printing fruit (lesson 3)
Class: Key stage 3, working between P1(i) on the QCA scale and National Curriculum level 1.

Starter – Recap on previous learning objectives

What are we learning about?

* Answers such as 'fruit' or names of individual fruits, or signs e.g. 'food' would also be an acceptable start.
* What can we do with fruit? 'printing', 'pressing' or some kind of gesture-imitating print action.
* Show the printing blocks used last week.
* Which part of the block makes the mark on the paper? 'Raised part' using gestures and pointing, etc.

Introduction and sharing of this lesson's learning objectives

This week you will learn:

* how to make a print block from the fruit you drew last week;
* the names of fruit we've all drawn (some will only learn their own fruit);
* what makes a good print block.

Alicia (who has more complex learning difficulties), I want you to be able to:

* taste two pieces of fruit;
* choose which you prefer;
* press a block which shows the fruit (made for you and supported by the TA).

Development

- Pupils work in small groups of two or three, supported by teaching assistants.
- Pupils decide which parts of the designs they drew last week to cut out of polystyrene.
- Pupils cut out the polystyrene and glue it to the wood blocks.
- When dry, they print with the blocks using a single colour.

Note levels of support required by individuals.

Plenary – Recap and evaluation as a group

- Who can tell me what they have learned today? Show or tell as appropriate to individual's ability. Encourage the use of key vocabulary.
- Evaluate prints, comparing them with the actual fruit and designs from previous lesson.
- Who can say which the fruit is? – name or match by pointing or eye-pointing.
- Does it look like we wanted it to? Identify reasons for smudging or other problems.
- Is there a difficulty in just using one colour?
- What could we do to overcome this? Make two blocks; add colour after with a brush; use another colour, etc.
- Next week we will use different ways of adding colour to make a two-colour print.

From: DfE, SEN: Training materials for the
foundation subjects (DfE Publications, 2003)

Appendix 6.6 Performance descriptions

Performance descriptions outline early learning and attainment before level 1 in eight levels, from P1 to P8.

Performance descriptions in D&T, P1–P3

P1 (i) Pupils encounter activities and experiences. They may be passive or resistant. They may show simple reflex responses, such as startling at sudden noises or movements. Any participation is fully prompted.

P1 (ii) Pupils show emerging awareness of activities and experiences. They may have periods when they appear alert and ready to focus their attention on certain people, events, objects or parts of objects, such as pausing over food smells in the room. They may give intermittent reactions, such as sometimes briefly grasping materials placed in their hands.

P2 (i) Pupils begin to respond consistently to familiar people, events and objects. They react to new activities and experiences, such as turning to a particular food item. They begin to show interest in people, events and objects, such as briefly focusing on the sound of a making activity. They accept and engage in coactive exploration, such as (with staff support) feeling the textures of wood, metal, plastic, fabric and foods.

P2 (ii) Pupils begin to be proactive in their interactions. They communicate consistent preferences and affective responses, such as turning towards a particular food item or a product that is their favourite colour. They recognise familiar people, events and objects, such as grasping the handle of a tool. They perform actions, often by trial and error, and they remember learned responses over short periods of time, such as lifting and lowering a tool or pressing their fingers into soft dough several times. They cooperate with shared exploration and supported participation, such as working with an adult to apply glue to a surface.

P3 (i) Pupils begin to communicate intentionally. They seek attention through eye contact, gestures or actions. They request events or activities, such as reaching out towards a particular piece of equipment. They participate in shared activities with less support. They sustain concentration for short periods. They explore materials in increasingly complex ways, such as tearing, squashing, mixing or bending materials. They observe the results of their own actions with interest, such as after bending sheet materials. They remember learned responses over extended periods, such as banging with a hammer.

P3 (ii) Pupils use emerging conventional communication. They greet known people and may initiate interactions and activities, such as pushing the spoon into the mixing bowl. They can remember learned responses over increasing periods of time and may anticipate known events, such as covering their ears before a loud sound. They may respond to options and choices with actions or gestures, such as picking up one tool rather than another. They actively explore objects and events for more extended periods, such as banging, scraping, rubbing or pressing tools against a surface. They apply potential solutions systematically to problems, such as pressing materials together.

From level P4–P8, many believe it is possible to describe pupils' performances in a way that indicates the emergence of skills, knowledge and understanding in D&T. The following descriptions provide an example of how this can be done.

Performance descriptions in D&T, P4–P8

P4 With help, pupils begin to assemble components provided for an activity, such as placing bricks together. They contribute to activities by coactively grasping and moving simple tools, such as a glue spreader. They explore options within a limited range of materials, such as adding grapes or chopped apple to a fruit salad.

P5 Pupils use a basic tool, with support, such as a roller. They demonstrate preferences for products, materials and ingredients, such as selecting a preferred filling for a sandwich.

P6 Pupils recognise familiar products and explore the different parts they are made from. They watch others using a basic tool and copy the actions, such as preparing a surface with a glasspaper block. They begin to offer responses to making activities, such as suggesting the colour or shape of a product.

P7 Pupils operate familiar products, with support, and explore how they work. They use basic tools or equipment in simple processes, chosen in negotiation with staff, such as in cutting or shaping materials. They begin to communicate preferences in their designing and making, such as adding selected felt shapes to fabric.

P8 Pupils explore familiar products and communicate views about them when prompted. With help, they manipulate a wider range of basic tools in making activities, such as joining components together to make their intended product. They begin to contribute to decisions about what they will do and how, such as communicating their approval of certain features of a process.

Appendix 7.1 How to make the most of your support staff

Practical tips:

- Set aside regular time to discuss the topic or lesson and your objectives with support staff. Prepare a brief scheme of work to show how learning could be developed if the assistant supports the student at other times (e.g. in a withdrawal lesson). Give staff a copy of the book (or books) to be worked from along with any background information on the topic.
- Help your support staff to become familiar with the room layout and organisation, setting up and clearing up routines (direct to department handbook for introduction, arrange for them to see the room when it is empty in order to locate equipment, books etc. Discuss how you organise the lesson(s) and the role that you expect them to take: e.g. who is in charge of different aspects of the lesson?
- Discuss the medical and/or learning difficulties of each pupil. Discuss pupils' individual education plans and support as appropriate.
- Discuss pupils who may need additional support at different stages throughout the project.
- Ask the assistant if they are familiar with the machinery, techniques and basic skills that will be used during the work. Indicate which pupils use special equipment or aids and how they are used by them.
- Identify health and safety issues and implications regarding them for the support team (e.g. safe use of machinery, possible hazards).
- Discuss time targets for lessons and when it will be appropriate to assist pupils to finish (without doing the work for them!).

Appendix 7.2 Daily planning grid

Class _____

Year group _____

Class teacher _____

TA _____

Week starting _____

Session			
To work with:			
Activity			
Particular focus			
Learning objective			
Things to look for			
Comments			
Odd jobs			
Notes			

Appendix 7.3 Weekly planning grid

Year group _____

Class teacher _____

TA _____

Week starting _____

	Monday	Tuesday	Wednesday	Thursday	Friday
Groups					
Activity					
Focus					
Learning objective					
Session after playtime					
Afternoon Session					

Theme for week _____

Appendix 7.4 Weekly support notes

Month _____ Name _____

	S1	S2		S3	S4
Monday			L		
Tuesday			U		
Wednesday			N		
Thursday			C		
Friday			H		

Key

S =

Session

Appendix 7.5 Individual support record

Pupil _____ Teaching Assistant _____

Date	Subject	Teacher	Objective/s	Responses

Appendix 7.6 Class support liaison record

Date	Subject	Focus pupils	Comments	Signed

Appendix 7.7 EHC plan targets and evaluation notes

Name _____

Objectives	Method	Period 1	Period 2	Period 3	Period 4
1.					
2.					
3.					
4.					

Appendix 7.8 Support diary

Week beginning: _____

Date	TA	Lesson	Comments

Week beginning: _____

Date	TA	Lesson	Comments

References and further information

DfE (2015) SEND Code of Practice. London: DfE Publications.

Lewis, A. (1992) 'From planning to practice', British Journal of Special Education, 19(1), 24–27.

DfE, SEN: Training materials for the foundation subjects (DfE Publications, 2003).

Ofsted (2004) Standards and Quality 2002/03 The Annual Report of Her Majesty's Chief Inspector of Schools.

Ofsted (2011) *Meeting technological challenges? Design and technology in schools 2007–10*. Available at: www.ofsted.gov.uk/publications/100121.

Sacks, O. *An Anthropologist on Mars: Seven Paradoxical Tales* (Sydney: Picador, 1995).

West, T. (2 Ed.) *In the Mind's Eye: Visual Thinkers, Gifted People with Learning Difficulties, Computer Imaging, and the Ironies of Creativity* (New York: Prometheus Books, 1997).

Websites

http://sirkenrobinson.com/pdf/allourfutures.pdf

https://dspace.lboro.ac.uk/dspace-jspui/handle/2134/2785

http://files.eric.ed.gov/fulltext/EJ960113.pdf

www.nicurriculum.org.uk/docs/inclusion_and_sen/gifted/gifted_children_060306.pdf

www.gov.uk/government/publications/below-the-radar-low-level-disruption-in-the-countrys-classrooms

Index

Page numbers in *italics* refer to figures.
Page numbers in **bold** refer to tables.